EQ For Engineers

Mastering Emotional Intelligence in a Logical

World

Nicci Brochard
&
Dr. Ben Chuba

EQ For Engineers

Mastering Emotional Intelligence in a Logical

World

CROSSBORDER

New York, London, Quebec

Contents

Chapter 4: Social Awareness – Empathy and Team Dynamics 74

Chapter 5: Relationship Management – Communication and Collaboration 106

Chapter 6: Leading with Emotional Intelligence – Guiding Teams with Heart and Logic 125

Chapter 7: Lifelong Growth – EQ in Your Career and Life 140

Introduction

Engineers solve impossible problems daily. You design bridges that span vast canyons, create algorithms that power global communications, and build systems that launch humanity into space. Your analytical mind dissects complexity into manageable components, your methodical approach transforms chaos into order, and your precision turns abstract concepts into tangible reality.

Yet despite these remarkable abilities, many engineers struggle with an equally complex system: human emotions. The same brilliant minds that can optimize code for maximum efficiency or calculate load-bearing requirements down to the decimal often find themselves puzzled by workplace dynamics, frustrated by unclear communication, or overwhelmed when leading diverse teams.

This disconnect stems from a fundamental misunderstanding. Emotions are not the opposite of logic, they are data. Just as you would never ignore critical sensor readings or dismiss important system metrics, emotional intelligence provides essential information for navigating the human elements of engineering work. Every breakthrough innovation, every successful project launch, and every career advancement depends not just on technical mastery, but on your ability to understand, manage, and leverage emotional dynamics.

The engineers who thrive in today's collaborative, fast-paced environment have discovered a powerful truth: emotional intelligence amplifies technical brilliance rather than competing with it. When you learn to read the emotional signals in a heated design review, manage your own stress during system failures, or inspire confidence in stakeholders during uncertain projects, your technical skills become exponentially more effective.

This book provides a systematic, evidence-based approach to developing emotional intelligence specifically tailored for the engineering mindset. You will discover practical frameworks, measurable techniques, and actionable strategies that treat emotional intelligence as another skill set to master. The future belongs to engineers who combine technical excellence with emotional fluency—and that future starts now.

Nicci and I (Ben) thank you immensely for choosing our book. We promise you a great time ahead.

Chapter 1

The Case for EQ in Engineering – Bridging Logic and Emotion

Engineers are often seen as ultra-logical problem solvers, people who live in a world of code, numbers, and precision. Yet even in this highly technical field, human factors and emotions play a pivotal role. Emotional Intelligence (EQ) – the ability to recognize, understand, and manage one's own emotions and influence others' emotions – is not a "soft" skill that's nice to have; it's a critical component of success. In this chapter, we'll debunk the myth that engineering is *only* about logic, explore how everyday interactions in engineering rely on empathy and communication, examine the human factors behind technical project failures, and highlight the tangible career benefits (the ROI) of developing EQ as an engineer. By bridging logic and emotion, engineers can enhance their problem-solving abilities, foster stronger teams, and propel their careers forward. Let's explore why EQ matters just as much as IQ in the engineering world.

Beyond the Stereotype – Emotions in a Technical Field

When you hear "engineer," you might picture someone hunched over a computer or blueprint, immersed in calculations and code. There's a long-standing stereotype that engineers are purely analytical beings – geniuses of logic who have little use for emotion. In fact, engineers are often stereotyped as highly analytical individuals, sometimes at the expense of social skills or emotional intelligence. This image implies that feelings and interpersonal insight don't factor into engineering work. But reality paints a different picture. Engineers are human, not robots, and emotions do influence their work, decisions, and effectiveness. Far from being a weakness, emotional intelligence complements analytical skills and can actually enhance problem-solving and innovation.

Think about the process of engineering design or troubleshooting. Beyond crunching numbers, it often requires creativity, imagination, and empathy. Why empathy? Because engineers ultimately design solutions *for people*. Whether it's software, infrastructure, or a consumer gadget, understanding the human context is crucial. Great engineers put themselves in the user's shoes and consider the needs and frustrations of real people. Research from educational leaders underscores this point: engineers are creative problem-solvers who must tap into empathy and creativity to consider a problem (and its solutions) from many different angles. In other words, an engineer who can empathize with end-users or colleagues can better identify the right problems to solve and devise innovative solutions that truly meet those needs.

Moreover, engineering challenges rarely have a single textbook answer. There are usually multiple approaches to any problem. Deciding on the "best" solution isn't purely a logical optimization exercise; it involves understanding trade-offs and perspectives. Emotional intelligence helps here by allowing an engineer to appreciate the viewpoints of various stakeholders – be it the concerns of a client, the ideas of a teammate, or the limitations of end-users. As one engineering professor put it, *"Good engineers are also empathetic... figuring out the best approach requires taking the time to understand the complexities of a problem from numerous perspectives."* Empathy fuels curiosity and open-mindedness, prompting engineers to ask the right questions: *Who will use this? What do they care about? How will this solution impact people?* By considering such questions, engineers escape the trap of tunnel-vision logic and often discover more innovative, user-friendly solutions.

Debunking the "pure logic" stereotype isn't about downplaying technical skill – it's about *enhancing* technical skill with emotional insight. Emotional intelligence can improve teamwork, creativity, and leadership. For example, an engineer with high EQ is more likely to remain calm under pressure, listen actively to a colleague's idea, or recognize when a project plan is causing team burnout. These aren't touchy-feely extras; they directly impact the quality of the work. An emotionally attuned engineer might realize that a teammate is frustrated with a complex module and offer help, preventing a small problem from festering into a major delay. Or consider innovation: a strictly analytical mindset might solve a problem *on paper*, but an empathetic mindset finds solutions that people embrace in practice. The myth that emotions have no place in engineering is fading as the industry sees how EQ skills actually amplify engineers' effectiveness. In short, being emotionally intelligent doesn't contradict being logical – it bridges logic and emotion, allowing engineers to apply their technical talents in the most impactful way.

Everyday Interactions, Extraordinary Impact

If you think an engineer's day is spent isolated in front of a screen, think again. Modern engineering is a team sport. On any given day, an engineer might brainstorm with a product manager, give or receive feedback on code, debate design decisions with a teammate, or negotiate timelines and requirements with a client. These "everyday" interactions are ordinary in occurrence but extraordinary in impact – they often determine whether projects run smoothly or fall apart, and whether teams thrive or struggle. The common thread in all these scenarios? Communication and empathy are key. In fact, one engineering blog notes that while developers love to "communicate through code," traditional communication skills (talking, writing, listening) are just as, if not more, important in day-to-day engineering work.

Let's break down a few routine engineer interactions to see how much they rely on EQ:

- **Giving and receiving code review feedback:** Code reviews are a staple of software engineering. When done with emotional intelligence, they become opportunities for learning and team bonding. When done without it, feelings can get hurt and team morale suffers. Imagine you submit code you've worked on for days. A teammate reviews it. If they respond with a blunt comment like "This is all wrong, what a mess," you're likely to feel defensive or demoralized. But if they approach it with empathy – perhaps saying, "I see what you were going for here. Maybe we can simplify this part; what do you think about trying X approach?" – the outcome is completely different. You feel respected and open to discussion. Showing empathy in how you communicate during code reviews leads to a more effective team by creating psychological safety (i.e. teammates feel safe to speak up and know that feedback comes from a place of care). In teams with high psychological safety, people aren't afraid to admit mistakes or ask questions, which means issues get surfaced and solved faster. In short, a little empathy in feedback goes a long way toward building trust and improving the quality of the code and the team.

- **Negotiating project requirements and timelines:** Engineers frequently work with non-engineers – clients, product managers, sales teams – to figure out what a product should do and when it can be delivered. These conversations can be fraught with emotion. The client wants all the features *yesterday*, the product manager pushes for a killer feature, and the engineering team is worried about unrealistic deadlines. Here, technical knowledge alone won't resolve the impasse. Emotional intelligence is crucial to understand each party's concerns and find common ground. An engineer with EQ will listen actively ("It sounds like timely delivery is your top priority; what if we prioritize these core features and discuss additions in the next phase?") and express their own constraints with empathy ("We want to deliver a great product for you, and we also don't want to over-promise and under-deliver. Let's work out a schedule that addresses your key needs within our capacity."). By reading the room, managing their own frustration, and showing they understand the other's perspective, an engineer can transform a tense negotiation into a collaborative problem-solving session. These soft skills often make or break not just the deal with the client, but the morale of the team implementing the project.

- **Team collaboration and conflict resolution:** Engineers often work in teams building complex systems, where components and people must interface smoothly. Misunderstandings happen – perhaps two developers implement connecting modules with mismatched assumptions, or there's disagreement on technical direction (say, which cloud service to use). Here, EQ is the oil that prevents friction from grinding progress to a halt. Recognizing emotions (am I or my teammate getting frustrated?), managing them (take a breather, communicate calmly), and empathizing (why does my colleague feel so strongly about this approach?) are steps that lead to constructive resolutions. Maybe your colleague is anxious about using a new technology and that's why they resist it – if you sense that, you might address the real concern (offer training or a pilot run) rather than arguing over merits in abstract. Everyday teamwork for engineers involves meeting tight deadlines, dealing with challenging relationships, coping with change and setbacks, and negotiating solutions with others. How we handle each of these situations is greatly improved by emotional intelligence. Teams with emotionally intelligent members tend to communicate better and have more mutual respect, which means they can coordinate complex tasks more effectively. It's often said that hard skills get you hired, but soft skills determine if you succeed – nowhere is that more evident than in the daily interactions of engineering work.

Take a moment to reflect on your own experience (or observation) in technical teams. Chances are, the projects that went well weren't the ones with zero bugs or genius-level code; they were the ones where team members clicked, information flowed freely, and everyone felt valued. And projects that went poorly often suffered from communication breakdowns or clashing egos. Those everyday chats, emails, code review comments, and meetings have an outsized impact. By building your EQ – being a bit more mindful of how you word that code review, how you listen in a meeting, or how you react to stress – you can dramatically improve not only the work atmosphere but also the final outcomes. "Soft" interactions often make or break team success, and emotional intelligence is what makes those interactions positive and productive.

When Technical Projects Fail – The Human Factor

Engineering is undoubtedly a technical endeavor – but when projects fail, it's usually *not* because someone couldn't solve a tough algorithm or debug a nasty error. It's because of people problems. Missed expectations, poor communication, lack of teamwork – these are the silent killers of technical projects. In fact, industry surveys and post-mortems often reveal that the root causes of failure are overwhelmingly human, not technical. One global developer survey found that software projects most often fail because of poorly documented requirements and a lack of effective communication. Think about that: not because the code didn't work, but because the team never fully understood what was needed or failed to talk to each other! Unclear goals, changing requirements without discussion, stakeholders not being on the same page – all are issues emotional intelligence could help prevent or mitigate.

Why Software Projects Fail. A survey of software professionals identified "changing or poorly documented requirements" (cited in 48% of failures) and "poor team or organizational management" (37%) as top reasons that projects derail. In contrast, purely technical factors (like inadequate tools or platform problems) were much further down the list. What do the top factors have in common? They're all about communication, understanding, and people. When requirements aren't clear, it often means the engineers and stakeholders weren't truly listening to each other or asking the right questions. When team management is poor, it usually reflects low trust, unresolved conflicts, or team members not feeling accountable – again, fundamentally human issues. Many of us have lived this: a project might have brilliant engineers onboard, but if they aren't communicating well or if there's constant friction, the project is likely to miss deadlines or fail outright. Miscommunication can cause a small bug fix to turn into a month-long fire drill because someone didn't feel comfortable admitting they didn't understand the specification. Poor teamwork can mean two subsystems don't integrate because their developers never ironed out who was responsible for what.

Emotional intelligence is the antidote to these human-factor failures. How so? Consider the requirement-gathering phase of a project. An emotionally intelligent engineer or project manager will practice active listening and empathy with the client or end-user. They'll pick up on not just *what* is being said, but *how* it's said – clarifying assumptions, noticing if a stakeholder seems uncertain or dissatisfied, and encouraging open communication. This can catch unclear requirements before they wreak havoc. During development, if something begins to go off-track (perhaps a misunderstanding about a feature), a team with good EQ will be able to have honest conversations without blame, thus addressing the issue early. Compare that to a low-EQ scenario where developers might hide problems out of fear or product owners might steamroll developers' concerns – the issue festers until it explodes.

Team dynamics are another critical piece. A technically brilliant team can fail if members don't collaborate. We've seen projects where one star programmer alienates others with harsh criticism, or where different departments point fingers rather than solve problems together. These are failures of emotional intelligence. On the other hand, teams that foster trust and respect – hallmarks of high EQ – can weather technical storms. If a major bug emerges close to a deadline, an EQ-driven team will pull together, communicate openly about what happened, and support each other to fix it. A low-EQ team might descend into panic and blame: developers hiding mistakes, managers shouting, everyone stressed and disengaged. The outcome in those two scenarios will be very different.

The evidence is so compelling that many engineering organizations now prioritize "team fit," communication skills, and adaptability just as much as coding ability. The reason is simple: a project is more likely to fail from neglecting the human element than from lacking technical expertise. By cultivating emotional intelligence, engineers can drastically reduce the risk of these failure modes. You start seeing potential communication pitfalls before they cause damage – for example, noticing that a junior developer looks confused and inviting them to speak up, or recognizing that a client's silence in a meeting might indicate unspoken concerns. In the logical world of engineering, EQ acts as a safety net, catching the *real* issues that logic alone might miss. The human factor can determine success or failure, so it's wise to give it as much attention as the technical details.

The ROI of Emotional Intelligence

Thus far we've made a case that EQ improves teamwork, innovation, and project outcomes – but does it really pay off in an engineer's career? Absolutely. Emotional intelligence isn't just a feel-good idea; it has very real professional and financial benefits. Think of EQ as an investment that yields high returns (ROI) in performance, leadership, and opportunity. Let's talk data: Studies have shown a strong correlation between emotional intelligence and top performance. Around 90% of top performers score high in emotional intelligence. That's a striking figure – it suggests that nearly all of the people who excel in their careers have strong interpersonal and self-management skills alongside their technical talents. It's not that they code better in isolation; it's that they leverage EQ to get better results. Additionally, employers have caught on to this fact. In one survey, 71% of employers said they value EQ more than IQ when hiring and promoting – a clear majority recognizing that teamwork, adaptability, and communication skills often trump raw intellect in the workplace. Why? Because an engineer with stellar EQ is likely to resolve conflicts, lead teams, and adapt to change, whereas an equally smart but low-EQ individual might cause rifts or struggle in leadership roles.

Emotional intelligence can even boost your paycheck. Research indicates that high-EQ professionals tend to earn more – potentially $29,000 more per year on average than their lower-EQ counterparts. Over a career, that's a massive difference. The reasons for this earning gap tie back to performance and leadership: people with higher EQ often climb into management roles, handle client relations more effectively, or excel in roles that require cross-functional collaboration – all of which are valued and compensated. In engineering, this might translate to being promoted from developer to team lead to engineering manager, or being entrusted with key client-facing responsibilities on a project. Your brilliant technical skills might get your foot in the door, but EQ is what opens the upper rungs of the ladder.

Don't just take it from the numbers – industry leaders are openly emphasizing emotional intelligence as a must-have. A prominent example comes from a chief operating officer in the engineering sector who was asked what advice he'd give young engineers. His answer: focus on EQ. He noted that emotional intelligence is "a critical skill" that will help engineering newcomers succeed in their careers and personal lives. This sentiment is echoed in many tech companies today. Engineering managers often say they can teach a new hire the specifics of a coding language, but they can't as easily teach them how to handle criticism or work under stress – that's on the individual to develop. Companies want team players and potential leaders. As an early-career engineer, if you demonstrate empathy, initiative in communication, and self-awareness, you're more likely to be given leadership opportunities or important projects because your managers see you can handle the responsibility. In contrast, the "brilliant jerk" engineer – the one who might produce great code but alienates teammates – often stalls in their career. In fact, some tech organizations have a policy of not hiring brilliant jerks at all, because they understand the cost to team morale and productivity.

The ROI of EQ also shows up in job satisfaction and well-being, which are important aspects of a sustainable career. High-EQ individuals tend to have better work relationships and handle stress more effectively, leading to a more positive work environment. This often means lower burnout and longer, more fulfilling careers – another payoff of the EQ investment, even if it's less easily quantified. From a business perspective, teams led by emotionally intelligent managers tend to perform better. They have lower turnover (people don't leave good managers who empathize and support them), and they innovate more, because team members feel safe to contribute ideas. All of these factors translate into a competitive edge for companies – and for you as an engineer within those companies.

In summary, developing emotional intelligence can be one of the smartest career moves an engineer makes. It boosts your individual performance (by helping you navigate workplace challenges smoothly), enhances your team's success (since you'll communicate and lead better), and opens doors to advancement (because EQ is a hallmark of effective leaders). It's telling that many employers now screen for EQ in interviews and some even provide training for it, knowing it leads to better outcomes. The logical world of engineering is, in truth, a very human world when it comes to building products and collaborating on teams. Mastering EQ gives you a powerful advantage: it's the skill that turns good engineers into great engineers, and technical experts into inspiring leaders. The data and industry voices make it clear – emotional intelligence isn't just a "nice-to-have," it's a critical asset with high ROI. By investing in your EQ, you're not only becoming a more well-rounded person, you're actively fueling your career growth and success in the engineering field.

As we conclude this chapter, remember that mastering emotional intelligence doesn't mean abandoning logic or downplaying your technical abilities. It means bridging the two – using self-awareness, empathy, and communication to amplify the impact of your logical mind. Early-career engineers, seasoned developers, and technical leaders alike can all benefit from stronger EQ. In the chapters ahead, we'll delve into specific EQ competencies and how to build them, but the case is already clear: EQ is a game-changer. In a field driven by logic, those who also harness the power of emotion will lead the way in innovation, teamwork, and success. The engineering world is evolving, and the most successful engineers will be those who master the human side of the equation as expertly as they do the technical. The logical and the emotional are not rivals in engineering – when combined, they make an unbeatable formula for excellence.

Chapter 2

Self-Awareness – Knowing Yourself in a Technical World

Understanding oneself is the cornerstone of emotional intelligence, even (or especially) in a logic-driven profession like engineering. Self-awareness gives engineers the ability to recognize their emotions, strengths, weaknesses, values, and motivations – providing a foundation for managing behavior and making conscious decisions. In a technical world that prizes rational problem-solving, knowing yourself is what allows you to channel your intellect most effectively. The following sections explore key aspects of self-awareness for engineers: from recognizing emotional triggers and biases, to honestly assessing your abilities, clarifying your core values, and using feedback to illuminate blind spots. Developing these facets of self-awareness will help you thrive both personally and professionally in engineering's logical arena.

Emotional Self-Awareness: Understanding Feelings in a Logical Environment

Engineers are trained to be rational, but that doesn't mean we operate free of emotion. Emotional self-awareness is about recognizing your own feelings and reactions in real time, even amid the hyper-logical atmosphere of technical work. This skill starts with identifying emotional triggers in your daily routine. For example, many engineers feel frustration surge during a tough debugging session or anxiety build up before a high-stakes presentation. These emotional responses are normal – what matters is noticing them and understanding what sparks them. Recognizing patterns in your emotional reactions is crucial. Do you often get irritable after hours of troubleshooting a complex bug? Does your heart race when you're about to explain your design to stakeholders? By observing these triggers, you gain power to manage them instead of being controlled by them. In fact, self-awareness of emotional triggers allows you to foresee when stress might cloud your thinking and take steps to stay composed, leading to better decisions under pressure.

Engineers can approach emotional self-awareness much like problem-solving. One recommended strategy is adopting a "debugging mindset" toward your own emotions: *treat emotional reactions like bugs – observe them, understand their triggers, but don't let them control the system.* In practice, this might mean when a build fails for the third time and frustration flares, you pause to mentally step back. Rather than slamming the keyboard or sending a heated message, you identify, "I'm feeling frustrated because the issue is persisting." This moment of recognition gives you a choice in how to respond next, instead of reacting on autopilot. Over time, noticing emotional cues – a quickened pulse, a flushed face, a sudden negative inner dialogue – acts like a monitoring system alerting you to high "emotional load." Self-aware engineers use these alerts to initiate coping strategies (take a break, refocus, seek help) before emotions overflow. As one software architect noted, understanding your own work patterns (like knowing you solve problems better when well-rested, or that unfamiliar code frustrates you) means you can adjust your approach and "work smarter, not harder". In short, emotional self-awareness lets you leverage emotions as data about your state, rather than ignoring them or being dominated by them.

Building emotional self-awareness can involve practical techniques familiar to engineers. Journaling is one useful tool: documenting your daily wins, setbacks, and feelings can help "recognize patterns in behavior" and pinpoint what triggers certain emotional reactions. For instance, you might discover that you consistently feel anxious on Monday mornings (perhaps due to the team meeting), or notice that a specific colleague's code reviews make you defensive. These insights are like diagnostics – once you see a pattern, you can address the root cause (maybe by preparing extra for the meeting, or improving communication with that colleague). Another technique is the "emotion pause" – taking a 10-second breath before responding to an upsetting email or a broken build, giving your brain a chance to engage rationally. This brief pause can prevent an emotionally-charged outburst and lead to a more measured response. By iterating on such habits, engineers train themselves to stay calm and think clearly, even when emotions run high.

The payoff for developing emotional self-awareness is significant. Engineers who understand their own feelings can achieve better self-control and decision-making in stressful situations. Rather than react impulsively to a provocation (snapping at a teammate or rushing a deployment out of frustration), a self-aware engineer recognizes the emotional spike and chooses a constructive response. Over time, this leads to a reputation for level-headedness. It also improves problem-solving: when you acknowledge emotions like frustration or burnout, you're more likely to step back and prevent mistakes, whereas ignoring those feelings can lead to poor decisions or tunnel vision. In essence, emotional self-awareness equips you to manage *your most complex system — yourself* — so that your logical mind and emotions work in harmony. It's the first step to mastering emotional intelligence in a technical world.

Authentic Self-Assessment: Owning Your Strengths and Weaknesses

Being self-aware also means having an accurate, authentic assessment of your own abilities. For engineers, this translates to understanding your strengths, acknowledging your weaknesses, and seeing your blind spots without letting ego or fear distort the picture. In a field that highly values competence, it can be tempting to overestimate your skills or hide your limitations. But true self-awareness requires dropping any pretenses and taking an honest inventory of what you do well and where you need improvement. Engineers who can candidly assess their strengths and weaknesses set themselves up for growth – they develop a *realistic confidence* grounded in actual abilities, rather than bravado. Just as importantly, they gain the respect of colleagues by demonstrating humility and trustworthiness.

Authentic self-assessment starts with recognizing your strengths: maybe you write very efficient algorithms, or you excel at debugging, or you have a knack for translating client requirements into technical specs. Knowing your strong suits helps you maximize your contributions and builds confidence. However, realistic confidence is not the same as arrogance – it coexists with an awareness of your limitations. Perhaps your documentation skills are subpar, or you struggle with public speaking, or you tend to procrastinate on writing tests. A self-aware engineer doesn't view these as shameful secrets but as areas to improve or compensate for. In fact, research shows that most people have plenty of "blind spots" in how they view their own competence – one study found 95% of people believe they're self-aware, yet only about 15% truly are. This means many professionals misjudge their strengths or weaknesses. By contrast, acknowledging what you don't know or can't do well is a mark of wisdom, not a weakness. It allows you to strategically fill those gaps – whether through learning new skills, partnering with others, or asking for help. As leadership experts note, leaving your pride at the door and recognizing that smart collaboration produces better results than trying to do everything alone is key to success in technical environments. In other words, knowing when to seek help or rely on teammates is itself a strength – it shows you're committed to the best outcome, not just your ego.

Consider how this plays out in practice. Imagine you're facing a problem outside your expertise – perhaps a database issue and you're a front-end specialist. An engineer lacking self-awareness might refuse to admit his weakness, spending days stuck or implementing a suboptimal fix rather than consult a database expert. A self-aware engineer would recognize the gap in knowledge and reach out to a backend teammate or mentor early. Far from losing credibility, this honesty tends to build credibility. Team members appreciate that you value their expertise and the project's success over personal pride. In fact, effective engineering teams foster a culture where asking for help is normalized and encouraged. Industry veterans observe that there's a persistent myth that "asking for help means you lack competence," but in reality *"nothing could be further from the truth. Seeking help isn't a sign of weakness – it's a sign of truly embracing strategy".* By tapping into others' knowledge, you often arrive at a solution faster and learn more along the way. Moreover, colleagues will trust you more because they know you won't let problems fester out of stubbornness. As one tech leader put it, *the best engineers aren't those with all the answers, but those who know where to find the answers – they ask, they listen, and they integrate others' input.*

Authentic self-assessment also involves continuous reflection and adjustment. It can be helpful to periodically take stock of your skill set as if performing a personal code review. What new strengths have you developed? What weaknesses have emerged or lingered? Being objective about your abilities might involve using tools or frameworks. Some engineers use self-assessment quizzes or skills matrices; others solicit input from peers (we'll discuss feedback in detail later). The goal is to align your self-perception with reality as closely as possible, because that alignment yields confidence that is *grounded*. When you truly know what you're good at, you can lead projects or volunteer for tasks with assurance – and when you know what you're not good at, you can approach those areas with humility and openness to learning. Counterintuitively, owning your weaknesses often makes you more credible, not less. It signals that you're self-aware and accountable, qualities every team values. Co-workers and managers are more likely to trust an engineer who says "I'm not very familiar with this tech, I might need some support" than one who overpromises and under-delivers due to unchecked blind spots.

In summary, authentic self-assessment means calibrating your ego to reality. It's recognizing that you, like everyone, are a work in progress. By identifying your strengths, you can deploy them for maximum impact; by admitting your weaknesses, you position yourself to address them and invite collaboration. This mindset creates a virtuous cycle: realistic self-knowledge leads to growth and better performance, which in turn reinforces confidence and humility. In the long run, engineers who practice honest self-assessment tend to advance further – they build "realistic confidence" that inspires others, and they avoid the pitfalls that befall those who are pretending to be something they're not. In a technical world, knowing the limits of your knowledge is as important as knowing the extent of it.

Core Values and Personal Drivers: Guiding Your Career with What Matters

Self-awareness isn't only about emotions and skills – it also means understanding what truly motivates you at a fundamental level. These are your core values and personal drivers: the principles, goals, and inner definitions of success that give your work meaning. In the rush of an engineering career, it's easy to get caught up in external metrics (salary, job titles, tech stack prestige) or lose sight of why you became an engineer in the first place. However, engineers with a firm grasp of their own values and drivers make more fulfilling career choices and remain more engaged in their work over the long run. Clarifying what matters most to you can act as a compass, ensuring that each job, project, or company you choose is aligned with your authentic self.

Start by asking yourself: *What do I value deeply in my work and life?* The answers will be personal – one engineer might prize innovation and creativity, another might value stability and work-life balance, yet another might be driven by a mission to help society through technology. For example, perhaps you feel most motivated when your work has a tangible positive impact on the community, or you thrive in environments where continuous learning is encouraged. Maybe you define success not just by promotions, but by developing open-source tools that others use, or by having time to explore hobbies outside work. There are no wrong answers, but it's crucial to identify your own. Studies on career fulfillment emphasize that aligning your work with your core beliefs and priorities is key to long-term satisfaction. In fact, values-based career decisions lead to greater job satisfaction, higher motivation, and lower risk of burnout. When your daily tasks resonate with what you find meaningful, you naturally feel more energized and committed. Conversely, if you take a job that clashes with your values (say, you value collaboration but the company culture is highly competitive and individualistic), you'll likely experience stress and disengagement over time.

Connecting with your personal drivers can guide major career moves. Consider an engineer who highly values entrepreneurship and autonomy – staying long-term at a bureaucratic corporation might slowly drain them, whereas joining a startup or launching their own product would light them up. Another engineer might value security and teamwork, making them happier at a stable large company with a close-knit team rather than a volatile startup, even if the startup offers more money. It comes down to *your definition of success.* If success for you means working on technology that aligns with your ethical beliefs, then choosing an employer whose mission you support becomes crucial. If success means achieving a certain lifestyle or family time, you might prioritize flexible employers or even a certain geographic location. An anecdote from a civil engineer, Caroline Haatveit, illustrates this point. Reflecting on her career decisions, she said: *"I wish I'd known the importance of making sure your career path aligns with your personality and your core values – what impact you want to make in society, and what type of work will make you happy and fulfilled."*. Early in her career, she took a job that wasn't a good values match and only later found one where the company culture and purpose resonated with her, leading to much greater satisfaction. Her story is a valuable reminder: when you find a workplace that "speaks to your personality," you'll just know – and it will accelerate your growth and fulfillment.

To leverage core values in your career, first clarify what your values are. This might involve some deep self-reflection since it's not always obvious. You could start by recalling times you felt extremely engaged or, conversely, very demotivated at work. What factors were at play? Perhaps working on a collaborative team made you feel alive, indicating teamwork is a core value; or you burned out doing meaningless maintenance tasks, suggesting you value impact or challenge. Some career advisors suggest making a list of values (integrity, creativity, financial gain, service, independence, etc.) and ranking them. Another approach is to define your personal mission statement – one or two lines that capture what you aim to contribute through your engineering career. This exercise can crystallize your drivers. For example, you might write, "As an engineer, I strive to *innovate sustainable solutions* that improve everyday life," or "My goal is to *lead teams* that build reliable systems and mentor new developers." Such statements highlight the motivations that should steer your choices.

Once your values and drivers are clear, use them as a filter for decisions. When evaluating a job offer or project, consciously ask: *Does this align with what truly matters to me?* An alignment doesn't mean every task is joyous, but overall the role should move you toward your personal definition of success. Aligning work with values has concrete benefits – research shows it makes your career more meaningful and sustainable. Engineers who integrate their core values into their goals tend to experience a sense of purpose, which fuels perseverance. They are also better at self-motivation: when you can connect a late-night coding session to a goal you deeply care about (say, building a product you believe in), it doesn't feel like drudgery in the same way. Additionally, having clear personal drivers helps in tough times. When you face setbacks or performance pressure, remembering *why* you do what you do (be it love of technology, desire to solve human problems, providing for family, etc.) can bolster your resilience. It's like an internal North Star – keeping you oriented amid the chaos of deadlines and bugs.

Finally, core values can guide career pivots and growth. The tech industry is broad; self-aware engineers navigate it by gravitating toward teams and projects that fit their values. For instance, if learning is a top value, you might choose roles that offer mentorship or involve new tech over those that are comfortable but stagnant. If freedom is a major driver, you might eventually opt for consulting or freelance work. There is no single path to success in engineering – you define success for yourself. By knowing yourself at this deeper level, you increase the odds of finding career satisfaction. You'll also avoid the common trap of chasing goals that aren't really yours (like pursuing management because it's expected, when what you actually love is coding technical problems). In summary, aligning your career with your core values and drivers isn't a touchy-feely notion; it's a strategic move for long-term engagement and success. When your work reflects who you are, you bring more of your energy and passion to it, and that often translates into better performance and innovation.

Feedback as a Mirror: Using Others' Perspectives to Reveal Blind Spots

Even the most introspective engineer has blind spots. Self-awareness is not developed in isolation – we need feedback from others as a mirror to see ourselves accurately. Research underscores this: people often overestimate their self-awareness, and external input is essential to close the gap. In fact, in a nearly five-year study, organizational psychologist Tasha Eurich found that while 95% of people think they're self-aware, only about 10–15% actually are in reality. This stark difference arises because we all have aspects of our behavior or impact that we don't fully see. We might think we're clear communicators, for example, while team members quietly find us confusing. Or we assume we're calmly managing stress when our body language seems tense to everyone else. Feedback acts as a reality check – a way to compare our self-perception with how others experience us. For engineers who pride themselves on objectivity, treating feedback as useful data on your "performance" as a teammate or leader can be especially helpful. It shines light on those blind spots so you can address them and continuously improve.

One effective method to gather this insight is 360-degree feedback, where you get input from a full circle of colleagues: peers, managers, direct reports (if you lead others), and sometimes even clients. Unlike a top-down review, 360 feedback provides a *holistic view* of your strengths and weaknesses as observed from different angles. This comprehensive approach is proven to uncover blind spots and increase self-awareness by presenting a more complete picture than self-assessment alone. When multiple coworkers echo the same observation – for instance, that you handle code reviews thoughtfully but sometimes struggle with deadline communication – it's hard to ignore. It validates an area you may need to work on or a strength to build upon. Experts advise taking the extra step to directly compare your self-assessment with the feedback from others. Where you rated yourself high but others saw you as average might indicate overconfidence or a blind spot; where you rated yourself low but others gave praise might reveal strengths you're underestimating. This exercise can be humbling, but it's incredibly enlightening. It shifts self-awareness from subjective introspection to something more evidence-based.

Even without a formal 360 program, you can cultivate the habit of seeking regular feedback from those around you. Think of it as extending your sensors beyond your own mind. Actively ask colleagues questions like, "How could I have communicated that design idea more clearly?" or "Is there anything I could improve in how I pair-programmed with you on that task?" When done sincerely, this shows others that you are open to growth, and it often elicits valuable pointers. Leadership coach Suzi McAlpine notes that participating in 360° reviews or other assessments provides useful data to "help you know and see yourself (and how others see you) more clearly". She also emphasizes frequently seeking critical feedback at work and looking for themes in it. If you hear similar comments from multiple sources, pay attention – that pattern likely points to a genuine blind spot or strength. For example, if three teammates in a row mention that you tend to interrupt in meetings, it's a sign to work on your listening skills. Or if several people appreciate your thorough testing habits, you might realize that's a leadership trait you can further leverage and perhaps mentor others in.

It's important to approach feedback with the right mindset. View feedback as a tool for improvement, not as personal criticism. This can be hard in the moment – nobody enjoys hearing negative remarks about themselves. However, engineers can remind themselves that feedback is essentially debugging your interpersonal performance. Just as you'd rather know about a bug in your code than have it silently cause issues, you should rather know about a behavioral bug (say, a tendency to micromanage or a lack of responsiveness) than have it hinder your team unknowingly. Maintaining a growth mindset is key: see each piece of feedback as an opportunity to update "Version You" to be even better. This openness will encourage more people to be honest with you, creating a positive feedback loop (pun intended) of improvement. Conversely, if you react defensively or ignore feedback, colleagues will shy away from giving it, and you'll remain in a bubble – one of the un-self-aware colleagues that can frustrate teams and even cut a team's chances of success in half. Indeed, lack of self-awareness in one team member can increase everyone else's stress and decrease overall morale. Recognizing this, the most effective engineers and leaders actively *pull* feedback rather than waiting for a crisis.

Using feedback as a mirror also means sometimes confronting uncomfortable truths and making changes. It's not enough to collect feedback; true self-awareness requires acting on it. If you learn that teammates find you dismissive of new ideas, practice deliberately inviting others to speak and acknowledging their contributions. If you discover people are "walking on eggshells" around your intense reactions to bugs, work on tempering your responses and reassuring others it's okay to bring up problems. It can help to create a simple improvement plan focusing on one or two key areas from feedback and checking in with a trusted colleague to gauge progress. Over time, as you address blind spots, you'll likely see a double benefit: your skills and relationships improve, and your external self-awareness grows, meaning you become more attuned to how others perceive you. This doesn't mean constantly worrying about others' opinions; it means having a realistic understanding of your impact on people and adjusting when needed.

In summary, no matter how self-reflective you are, you can't see the whole picture of "you" without reflection from others. Feedback serves as that mirror, showing you angles you'd otherwise miss. Embracing feedback – whether through formal 360 reviews or informal conversations – is a powerful accelerator for an engineer's personal development. It uncovers hidden weaknesses you can strengthen and hidden strengths you can capitalize on. By actively seeking and utilizing feedback, you demonstrate humility and a commitment to growth that colleagues will respect. More importantly, you ensure that your self-awareness isn't an illusion but rather a continually sharpening image of who you are. In the technical world, this complete self-knowledge will enable you to lead yourself and others more effectively, making you not just a smarter engineer, but a wiser one.

Overall, self-awareness is the keystone of emotional intelligence for engineers. It enables everything else – self-regulation, empathy, communication – by giving you an accurate internal compass. By understanding your emotions and triggers, you gain self-control in high-pressure moments. By realistically assessing your strengths and weaknesses, you build confidence and credibility. By knowing your core values, you steer your career toward satisfying destinations. And by welcoming feedback from others, you correct your course and continue growing. In a field dominated by logic, cultivating these aspects of self-awareness will distinguish you as an engineer who not only thinks critically, but also understands themselves deeply – a combination that truly drives success in a technical world.

Chapter 3

Self-Management – Staying Cool Under Pressure

Anyone can tell you that in the high-pressure world of engineering, staying cool under pressure is as crucial as technical skill. Self-management – the ability to regulate your emotions, thoughts, and actions – is what keeps you composed when servers crash, deadlines loom, or the prototype fails for the tenth time. It's about handling stress and setbacks with resilience, calm, and optimism. This chapter explores how engineers can bounce back from failures, keep their composure in chaos, stay motivated through challenges, and maintain a healthy work-life balance. The tone here is conversational and supportive, with real-life examples, research insights, and case studies to inspire you. Whether you're debugging in a midnight outage or pushing the boundaries of innovation, mastering self-management will help you respond thoughtfully instead of reacting impulsively. Let's dive into the key aspects: resilience under pressure, emotional regulation and calm, staying motivated and positive, and balancing work with well-being.

Resilience Under Pressure: Bouncing Back from Setbacks

Imagine you've been troubleshooting a complex bug for days. You've tried fix after fix, and nothing works. Frustration builds with each failed attempt. At this point, many people feel the urge to give up – but resilience is what keeps you pressing on. In engineering, failure isn't just common; it's often an essential part of eventual success. Solving complex problems means failing multiple times and learning from each failure. Sometimes the breakthrough only comes on the 11th attempt (or the 40th!). For example, the water-repellent product WD-40 got its name because the first 39 formulations didn't work – the successful formula came on attempt number 40. Likewise, Sir James Dyson built 5,127 prototypes of his revolutionary vacuum design before finally getting it right. These innovators treated each setback as a learning opportunity. Every misstep yielded new information, guiding them closer to a solution. As Dyson later said, those thousands of "failures" taught him invaluable lessons about what didn't work, which made the eventual solution possible.

Resilience under pressure is the ability to bounce back and adapt in the face of difficulties. It's knowing that a failed test or a design flaw isn't the end of the road, just a hurdle to overcome. As one engineering professor put it, *"What are you going to do when you fail multiple times trying to solve a problem? It may be tempting to quit by the tenth try — but what if the solution comes on the eleventh?"* The best engineers internalize this mindset. They persist through challenges with the belief that each attempt gets them closer to success. Thomas Edison captured it well when he famously said, *"I have not failed. I've just found 10,000 ways that won't work."* That attitude of optimistic persistence is at the heart of resilience.

Strategies for building resilience: Start by reframing failures as feedback. Instead of thinking "This is a disaster, I'll never fix it," tell yourself "Now I know one more approach that doesn't work – I'm narrowing down the solution." This mental shift – often called a *growth mindset* – allows you to see value in mistakes. Each setback is simply data. Ask yourself: *"What can I learn from this? How will it inform my next attempt?"* By treating failures as learning opportunities, you maintain optimism and forward momentum.

Another key strategy is to maintain perspective. Under pressure, it's easy to catastrophize – to feel like one bug or blown presentation will ruin everything. Resilient engineers step back and view the bigger picture. Yes, this one attempt failed, but that doesn't mean *I* am a failure or that the goal is impossible. Remember times you overcame other problems; it reinforces that you're capable of solving this one too. Keeping an eye on long-term goals can fuel perseverance in the short term. For instance, if your circuit design keeps overheating, remind yourself that mastering this challenge will make you a better engineer and that the end product (say, a more efficient device) will be worth the struggle.

Real-world example – SpaceX's rocket failures: A striking case study in engineering resilience is SpaceX's journey to landing reusable rockets. Early on, SpaceX engineers watched several multi-million-dollar rockets explode or miss their landing barges. It would have been easy to see those crashes as insurmountable failures. Instead, the team treated each crash as a chance to improve the design and algorithms. They combed through the telemetry data, understood what went wrong (fuel depletion, guidance system errors, etc.), and made adjustments for the next test. There was a string of failures, but the engineers persisted with a mindset of *"failure is not final; it's data."* Eventually, after many attempts, they achieved the first successful vertical rocket landing – a breakthrough that only happened because they didn't give up after the earlier failures. This resilience under intense public pressure paid off, leading to technology that is now revolutionizing space travel. The lesson for everyday engineers is that persistence and adaptability in the face of setbacks can yield groundbreaking results, whether you're debugging code or building rockets.

In summary, resilience means staying in the game. It's bouncing back from each setback with renewed determination. You can cultivate resilience by celebrating small progress (even if the main problem isn't solved yet), keeping a positive outlook that the solution *will* emerge, and learning actively from every attempt. Engineering breakthroughs often require many iterations – your resilience ensures you'll still be there, trying again, when attempt number 11 (or 111) finally succeeds.

Emotional Regulation and Calm: Keeping Composure in High-Stress Moments

Pressure is part of an engineer's life: the site is down at 2 AM, a critical demo is hours away, or a meeting grows heated over a design review. In these moments, emotional regulation – the art of managing your emotions and staying calm under stress – is an invaluable skill. Teams and projects fare much better when someone can keep their cool and think clearly, instead of letting panic or anger take over. A calm mindset in a crisis can be the difference between a thoughtful solution and a costly mistake.

Consider a real-life scenario: A systems engineer at a startup accidentally deleted a crucial storage volume during a routine update, instantly knocking out the company's email servers. Suddenly, people were hovering around his desk, alarm bells ringing – a truly high-stress moment. His vision blurred and heart raced as panic set in; it's a natural fight-or-flight response when things go terribly wrong. But here's where emotional regulation came in: he recognized that he was panicking and deliberately hit "pause." He stood up, excused himself for a minute, and went to splash cold water on his face and take a few deep breaths. This simple break interrupted the adrenaline surge and cleared his mind. In that brief calm, he remembered a critical detail – he had set up a backup mirror of the storage volume earlier. The data was not lost after all. Returning to his desk with composure, he quickly reattached the backup and restored service. Within half an hour, the crisis was resolved. Had he remained in blind panic, he might not have recalled the backup or could have made the situation worse by reacting impulsively. By managing his emotions, he saved the day.

This story shows that even when your brain screams "panic!", you can take actions to regain calm. Techniques for keeping composure start with the basics: breathing and grounding yourself. When stress spikes – your code isn't working and the deadline is in an hour – your heart rate and blood pressure shoot up as the body responds like it's facing danger. Simply taking slow, deep breaths can counteract this. It activates your body's calming response (the parasympathetic nervous system), slowing the heart rate and helping clear the fog of anxiety. Many engineers find that pausing to inhale deeply (count to four), then exhale slowly (count to four) for a few cycles can significantly steady their nerves in a tense moment. It sounds almost too simple, but in the heat of the moment we often forget to breathe properly.

Another quick technique is to take a short break if possible. Step away from the problem for five minutes – get a glass of water, do a few stretches or a brisk walk down the hall. This isn't avoiding the issue; it's resetting your mind. Often, the moments when you feel "I have no time to step away!" are exactly when a brief pause is most valuable. As one Yale emotional intelligence researcher noted, our creative brain regions only activate when we're at ease – not when we're in full panic mode. In fact, engineers at Stanford found that a 15-minute break for mindfulness (like meditation or a quiet walk) boosted their creative thinking and led to more innovative solutions once they returned to the problem. So if you're stuck or freaking out, doing something as simple as walking around the block or splashing water on your face can literally help your brain reboot. You might come back with a fresh perspective or remember something crucial (just as that systems engineer recalled his backup).

Managing negative emotions like anger or frustration is also key to staying calm. Let's say in a meeting a teammate unexpectedly blames your module for a project delay, and you feel anger flaring up. Your first impulse might be to snap back defensively. But emotional intelligence means *not* reacting on impulse. One strategy is to name the emotion to yourself ("I'm feeling really angry and embarrassed right now"). It might sound odd, but naming what you feel actually helps reduce the intensity of that feeling – it gives your logical brain a moment to catch up to your emotional brain. With that small gap, you can choose a more constructive response. Perhaps you take a slow breath and respond, "I understand you're frustrated. Let's analyze the timeline issue together after this meeting." By staying composed and respectful, you prevent an escalation and demonstrate leadership under pressure. The ability to manage anger, anxiety, or frustration in real time means you respond thoughtfully rather than letting the emotion control you.

There are also longer-term habits that strengthen your emotional regulation. Regular mindfulness meditation, for example, trains your mind to observe feelings without being overwhelmed by them. Some engineers use meditation apps or simple breathing exercises daily, which can translate into calmer responses during actual crises. Even without formal meditation, you can practice awareness: next time you feel stress rising – a knot in your stomach during a tough debug – consciously remind yourself to "focus on the facts, not the fear." Replace inner statements like "I can't handle this" with "One step at a time – what's the next thing I *can* do?" This kind of positive self-talk keeps you solution-oriented.

Why staying calm matters: Engineering teams greatly value someone who can remain cool and collected when things go awry. If you've ever been in a late-night troubleshooting war room, you know how chaos can spiral if everyone is freaking out or blaming each other. A person who says "Alright, let's work the problem" in a steady voice can anchor the whole group. Calm is contagious – your composed demeanor can help others settle down, leading to a more focused team effort. On the flip side, unchecked panic or anger is also contagious and can derail the team. That's why companies seek out engineers with steady temperaments for leadership roles. The ability to stay calm and controlled in tough situations is highly valued in engineering teams, because it leads to better outcomes under pressure. When you keep your head, you make smarter decisions, catch errors, and guide others effectively. It turns a potential disaster into a challenge that can be managed. By practicing emotional regulation, you'll not only solve high-stress problems more effectively, but you'll also earn a reputation as the "cool-headed" colleague people trust when the stakes are high.

Staying Motivated and Positive: Cultivating Perseverance and Optimism

Engineering projects can be marathons, not sprints. There are times when the work becomes tedious, the obstacles mount, or progress seems painfully slow. Maintaining self-motivation and a positive mindset in these moments is crucial to avoid burnout and keep pushing forward. Emotional intelligence plays a big role here – it helps you harness internal motivation (beyond just a paycheck or fear of a boss) and cultivate optimism even when the going gets tough. This section explores how to set yourself up to stay energized and hopeful throughout your engineering journey.

The power of a positive mindset: Optimism isn't about wearing rose-colored glasses or ignoring problems. It's about believing that challenges can be overcome and that your effort will pay off. In a demanding work environment, a positive mindset fuels perseverance. For example, consider a software engineer debugging a complex system crash. A pessimistic outlook might lead them to think, "This is impossible, I'll never figure it out," which pretty quickly saps motivation to even try hard. In contrast, an optimistic engineer would think, "This is tough, but if I methodically check each module, I *will* find the cause." That belief in a solution keeps them engaged and pushing on. Indeed, research in psychology has found that optimistic individuals tend to be more resilient in the face of stress – they persist longer on difficult tasks and handle setbacks better. In engineering, that can translate to finding the critical bug on the 12th test rather than giving up after the 5th.

One way to stay positive is to reframe negative thoughts when they pop up. We all have an inner dialogue, especially under stress. Pay attention to it. If you catch thoughts like "I'm not good enough for this project" or "Everything is going wrong," challenge them. Replace them with more constructive frames: "This project is a chance to grow my skills," or "Some things are going wrong, but what can I set right?" Over time, this practice of reframing trains you to approach problems with a solution-oriented attitude rather than a defeatist one. It's essentially applying engineering thinking to your mindset: identify the unproductive "thought bugs" and refactor them into better ones.

Self-motivation strategies: Highly effective engineers often set *personal challenges* that spark their motivation. For instance, if you're working on a dull-but-necessary component, you might turn it into a game for yourself – "Can I refactor this module in half a day?" or "Let's see if I can reduce the memory footprint by 20%." Setting small goals or personal milestones within your work can create a sense of progress and achievement. Each time you hit one of those mini-goals, take a moment to acknowledge it – celebrate small wins. This could be as simple as marking a task done on your to-do list (that little checkmark can feel oddly satisfying!), sharing the news with a colleague, or rewarding yourself with a coffee break. Celebrating small wins isn't trivial; it has a powerful psychological effect. Accomplishment releases dopamine in the brain, giving you a feel-good boost that reinforces your enthusiasm. As management experts note, consistent progress – even small steps – is one of the biggest motivators for people in creative and analytical fields. So break big challenges into bite-sized tasks and then relish the completion of each bite.

Another tactic to maintain motivation is connecting your work to a bigger purpose. It's easy to lose drive if you feel like you're just churning out code for no reason. Take a step back and consider: *How does my work impact others or contribute to a larger goal?* Maybe your code is part of an application that helps doctors manage patients, or your design will make airplanes safer. Reminding yourself of the meaning behind the work can rekindle passion on days you feel drudgery. Many engineers find motivation in knowing their efforts matter – that there's a real-world benefit or a user's problem being solved. If the direct purpose isn't obvious, set a personal purpose: "I'm learning a lot through this project which will make me a better engineer," or "Completing this will open doors for more exciting work later." Purpose is fuel for perseverance.

It's also important to nurture an environment of positivity around you. Emotions are contagious, so try to surround yourself (as much as possible) with teammates who have upbeat, can-do attitudes. If you're a team lead, you can set this tone by acknowledging efforts, encouraging one another, and not dwelling on mistakes beyond learning from them. Something as small as starting stand-up meetings with each person sharing a "win" or something good can lift the team's mood and motivation. As an individual, you can influence the vibe by being the person who says "Alright, that test failed, but we got further this time – let's troubleshoot the next part." Positivity doesn't mean ignoring problems; it means *choosing* a hopeful approach to solving them.

Avoiding burnout through emotional intelligence: Even with optimism, engineering work can grind you down if you never come up for air. Self-management means noticing when your fuel gauge is low and taking action before you burn out. One aspect is to practice self-compassion – don't beat yourself up for feeling unmotivated at times. It happens to everyone. Instead, use EQ to investigate why: Are you exhausted? Do you need to ask for help or resources? Perhaps you've been focusing on what's going wrong and need to remind yourself of past successes. A positive mindset includes being kind to yourself. Think about how you'd encourage a friend in your situation, and apply that to your own self-talk.

Interestingly, helping others can boost your own motivation too. If you're stuck in a rut, try mentoring a junior engineer or assisting a colleague with their problem. Sharing advice or teaching someone can reignite your confidence and passion – it reminds you of how far you've come and why you enjoy engineering in the first place. Studies have shown that giving advice can increase the advisor's own motivation because it bolsters their sense of competence and purpose. So, by lifting someone else up, you often lift yourself up as well. It creates a positive feedback loop of enthusiasm.

Lastly, when you do hit a milestone or finish a tough project, acknowledge the achievement before rushing to the next thing. Perhaps take the team out for lunch or simply reflect for a moment on what you're proud of. Savoring achievements builds a reservoir of positive feelings you can draw on when the next challenge arrives. It reinforces the narrative that difficulties can be overcome and that your efforts lead to success. Armed with that confidence, you'll stay motivated even when you hit the next rough patch, because you know from experience that perseverance pays off.

In sum, staying motivated and positive in engineering is about managing your inner drive and outlook. Set yourself up with small goals, recognize your progress, keep your self-talk hopeful, and find the deeper meaning in what you do. By doing so, you maintain momentum and enthusiasm for your projects, even when the work gets difficult. You'll find you're not only more productive, but you also enjoy the journey more, turning challenges that could be dreary into opportunities to grow and triumph.

Work–Life Balance and Well-Being: Recharging to Sustain Success

Self-management isn't just something you practice at work – it extends to how you manage your life outside of work, too. In fact, what you do off the clock greatly affects your performance on the clock. Engineering is demanding mental work, and without proper rest and balance, even the most passionate engineer can fall prey to burnout. That's why this section focuses on maintaining a healthy work–life balance and personal well-being. It might seem tangential to career success, but consider this: if your mind and body are running on empty, how effective can you be in solving complex problems? Taking care of yourself is taking care of your career.

The importance of recharging: Think of yourself as a high-performance engine – you can't run it at redline constantly without breaking down. Regular maintenance and cool-down periods are necessary. Many of us have experienced or seen the effects of poor balance: an engineer pulling successive all-nighters ends up coding sloppy bugs, or a colleague who hasn't taken a vacation in years suddenly loses motivation or falls ill. These are signs that ignoring self-care undermines productivity in the long run. On the flip side, engineers who make time to recharge tend to have better focus, more creativity, and sustained productivity. There's a reason top tech companies have relaxation pods, game rooms, or gym facilities on campus – they know that a balanced, refreshed mind is more effective than an overworked, frazzled one.

Healthy routines to manage stress: Start with small daily habits. For example, take regular breaks during the workday. It's easy to get lost in code for hours, but short breaks actually improve concentration. Every hour or two, stand up and stretch or walk around for a few minutes. Maybe grab a healthy snack or chat briefly with a coworker. These mini-breaks prevent stress from accumulating and often you return with a clearer head. As an engineer, I've often had the experience where stepping away from the screen for five minutes leads to an "aha!" moment on a bug that seemed unsolvable. It's like giving your brain a short nap – it comes back sharper.

Another routine is mindfulness or relaxation practices. This could be as simple as a 10-minute meditation in the morning, or just a quiet ritual like drinking tea on the balcony after work. Some engineers practice mindfulness meditation to reduce anxiety and improve concentration. If meditation isn't your thing, no worries – find what relaxes you, whether it's listening to music, journaling, or even playing a calming video game for a bit. The goal is to regularly downshift your brain from high gear to a lower gear, so that you're not constantly in a state of tension. This makes your "up time" more effective. Think of it as intervals of stress and recovery, much like physical exercise.

Physical well-being is just as crucial. We sometimes forget that the brain is part of the body, and a healthy body supports a high-performing brain. Regular exercise is a proven stress reducer – it pumps up endorphins (natural mood lifters), improves sleep quality, and can even enhance cognitive function. You don't have to become a marathon runner; even a 30-minute brisk walk or a bike ride a few times a week can do wonders. Many engineers find that some of their best ideas or insights come during a jog or while shooting hoops, because exercise clears the mind and often puts you in a kind of reflective mental state. Plus, staying fit boosts your energy levels during those long workdays.

Setting boundaries between work and personal life: With remote work and always-connected culture, many of us struggle to truly unplug. But setting some boundaries is vital. For instance, you might set a rule to not check work email after 8 PM, or designate one weekend day as completely work-free. If you're in a crunch period and extra hours are needed, balance it out when the crunch is over by taking some comp time or lighter days. Communicate your boundaries to your team if needed – most managers today understand the importance of work-life balance (and if they don't, you can politely educate them or set an example). It might feel like "if I work an extra 2 hours tonight, I'll get more done," but if those 2 hours come at the cost of your sleep and sanity, you may pay back that debt with interest in lost productivity later. Working smarter, not just longer, is the key. And part of "smarter" is knowing when to call it a day.

Recharging outside of work: Make sure to engage in hobbies and social activities that fulfill you. Whether it's playing guitar, cooking, hiking, or spending time with family and friends, these activities replenish your mental and emotional reserves. They might seem unrelated to your career, but they are not. A well-rounded life gives you more mental resilience. For example, an engineer who has a fun weekend with friends comes back Monday re-energized and in a better headspace to tackle problems, compared to one who spent the weekend anxiously checking code repositories. Hobbies also provide a sense of accomplishment and joy outside of work, which can protect your self-esteem from being entirely tied to your job successes or failures.

Avoiding burnout: Burnout is a state of chronic stress and disengagement – it creeps up when there's been too much pressure for too long without sufficient recovery. One way to prevent burnout is to regularly assess your stress levels and act before you hit a breaking point. Pay attention to signs like constant exhaustion, cynicism (e.g., feeling "what's the point" about work you used to care about), or declining performance. These are warning lights on your dashboard. If you see them, don't ignore them – take action. This might mean taking a few days off, talking to a manager about adjusting workload, or even seeking support from a mentor or counselor. High emotional intelligence includes recognizing when you need help or a change, rather than just soldiering on until you collapse.

Companies are increasingly understanding that well-being enhances productivity. Studies show that employees with a good work-life balance tend to be more engaged and productive during work hours. When you're rested and happy, your focus is sharper and you can achieve more in less time. It's not just feel-good rhetoric – there's data suggesting that, for example, working 55+ hours a week for long periods actually yields diminishing returns and more errors compared to a balanced 40-hour week with proper rest. Think of the times you solved a problem in 1 hour fresh in the morning that you struggled with for 4 hours the night before when you were exhausted. That's the work-life balance argument in a nutshell.

Case in point: A hardware engineering team at a tech firm found themselves in perpetual "crunch mode," working late nights for months. Output initially was high, but soon bugs started slipping into designs and team morale plummeted. Recognizing the unsustainable pattern, management enforced a new rule: no emails after dinner time, and at least one day off on weekends. They also organized optional group activities like Friday afternoon sports and encouraged taking vacation days. At first, the team worried productivity would drop. Instead, the opposite happened – refreshed engineers came up with creative solutions faster, and the product quality improved. The project still hit its deadline, but with far fewer hiccups and a much happier team. This turnaround underlines a simple truth: maintaining your well-being isn't a luxury; it's part of doing your best work.

Practical tips for balancing and well-being:

- **Prioritize sleep:** It's tempting to sacrifice sleep to squeeze in more work (or more play), but adequate sleep is non-negotiable for cognitive function. Problems that seem unsolvable at midnight often become clear after a good night's rest because your brain consolidates memories and ideas during sleep.

- **Take your vacations (and truly disconnect):** Vacations and days off exist for a reason. Don't be the person who has a stack of unused vacation days or who "vacations" with their laptop in tow. Taking time off to travel, or even a staycation to relax at home, can reset your stress levels. When you return, you'll likely find your motivation and focus renewed.

- **Set aside "unplugged" time daily:** Have at least an hour before bed (or another convenient time) where you do no work and preferably stay away from screens. Read a book, talk to loved ones, or meditate. This helps your mind unwind and signals that work is done for the day.

- **Healthy body, healthy mind:** We mentioned exercise, but also pay attention to nutrition. Skipping meals or living on caffeine and vending machine snacks will exacerbate stress. Try to eat regular, balanced meals – your brain needs steady fuel. And remember to hydrate; even mild dehydration can affect concentration.

- **Create transition rituals:** If working from home, it can be hard to separate work and personal life. Create a ritual to "end" your workday – maybe a quick walk outside when you log off, or changing into different clothes, or doing a short workout. This cues your brain that it's now personal time, helping you mentally disengage from work matters for the evening.

By building these habits and boundaries, you protect yourself from chronic stress. You'll find you have more energy and enthusiasm not just at work, but in life overall.

And here's the real win-win: when you do sit down to work, you'll be more focused, efficient, and creative because you're coming from a place of balance.

Mastering self-management is like adding a secret weapon to your engineering toolkit. With resilience, you won't be deterred by setbacks – you'll use them as stepping stones. With emotional regulation, you'll navigate crises with a clear head and steady hand, earning trust and achieving better outcomes under pressure. With a positive, self-motivated mindset, you'll push through challenges and continue growing, rather than stagnating or burning out. And by caring for your work-life balance and well-being, you ensure that you can perform at your best consistently, for the long haul. Emotional intelligence might sound "soft," but as we've seen, it has hard benefits: higher productivity, more innovation, stronger teamwork, and a more rewarding career.

As an engineer in a logical world, cultivating these emotional skills will set you apart. You'll be the one who stays cool when others panic, who perseveres when others quit, who motivates the team with optimism, and who lasts in a demanding field because you take care of yourself. It's both about thriving in your career and enjoying it. So the next time you face a high-pressure situation, remember these lessons. Take that deep breath, reframe the challenge, tap into your resilience, and know that you've got the tools to handle it. In doing so, you'll not only solve the problem at hand – you'll also lead by example, showing that emotional intelligence is indeed a superpower for engineers in a logical world. Keep cool, keep going, and success will follow.

Chapter 4

Social Awareness – Empathy and Team Dynamics

Understanding and navigating the human element in engineering is just as critical as mastering the technical aspects. In this chapter, we explore how emotional intelligence—especially social awareness—can transform an engineer's effectiveness in a team setting. Social awareness means tuning into others: recognizing subtle cues, empathizing with teammates and users, embracing diverse perspectives, and reading the room (or the entire organization) to know how best to contribute. These skills may not be taught in engineering school, but they prove invaluable in detecting issues early, fostering trust, and ultimately leading teams to better outcomes. Let's delve into four key facets of social awareness: active listening beyond words, empathy in action, diversity in thought, and reading team and organizational culture.

Listening Beyond Words: Developing the Art of Active Listening and Observation

Engineers pride themselves on logical thinking, but communication is often where projects succeed or fail. "Active listening" is the practice of fully focusing on the speaker, observing not just their words but also their tone and body language, and providing feedback to ensure understanding. It's about listening beyond the literal words being said. Studies have shown that only a small fraction of communication is in the actual words we use – one famous UCLA study attributed as little as *7% of communication to words*, with tone of voice and body language conveying the rest. While the exact numbers can vary by context, the lesson is clear: an engineer must pay attention to *how* something is said and what is not being said. For instance, a teammate might say "I'm fine" but avoid eye contact and speak in a flat, subdued tone – strong signs that things are *not* fine. Listening beyond the words in such a case means noticing the hesitation or the forced smile that accompanies the statement.

Active listening involves giving your undivided attention and signaling to others that you value their input. This can be as simple as nodding in agreement, maintaining eye contact, or saying "Hmm, I see" at appropriate moments. These nonverbal cues show the speaker that you're engaged. In fact, research confirms that positive nonverbal feedback – like nodding or an open posture – makes people feel heard and respected, which boosts trust and morale. An aerospace engineer we spoke to described how he makes a point of closing his laptop and turning fully toward colleagues during design discussions. *"When I give them my full attention – no screens, no distractions – they know I actually want to hear their ideas. It builds respect,"* he noted. By simply listening intently, he often uncovers concerns early in the development process. Team members become more willing to voice a subtle worry about a requirement or a potential design flaw. Catching those issues when they're still just worries, rather than full-blown problems, can save countless hours later on. As one leadership expert put it, *"active listening helps you build trust and understand other people's situations and feelings"*. In turn, the speaker feels valued and safe to share more, creating a virtuous cycle of openness.

Active listening also means paying attention to the team's morale in everyday interactions. Engineers are trained to focus on data and facts, but it's equally important to gauge the "data" coming from human signals. Is the team unusually quiet in meetings this week? Do you hear sighs or see slumped shoulders when a new deadline is announced? These observations can clue you in to stress or frustration levels in the group. For example, consider a software project meeting where the project manager asks if anyone has concerns. If nobody speaks up but you notice people exchanging glances or sitting with arms crossed, an *active listener* will sense the unspoken tension. Perhaps the timeline is too tight or requirements are unclear, but the team is hesitant to voice it. By recognizing these nonverbal cues, a socially aware engineer might gently say, "I have a feeling some of us might be worried about the schedule – shall we discuss the risks openly?" This invitation, prompted by listening between the lines, can bring hidden issues to light. Indeed, business communication research suggests that effective listening allows leaders to *identify potential issues early* and maintain team morale by addressing concerns before they escalate.

Another aspect of listening beyond words is observation – sometimes the most important information is what colleagues *aren't* saying. An example from a real engineering team comes to mind: A senior developer noticed that a usually vocal junior engineer had gone mostly silent during technical planning sessions. Instead of ignoring this change, the senior engineer reached out one afternoon for a casual chat. It turned out the junior engineer had reservations about the chosen technical approach but felt intimidated and didn't want to contradict the team lead. By listening to the *absence* of the junior's normal input and kindly inquiring, the senior engineer uncovered a critical concern. They brought the junior's perspective into the next meeting (without singling them out) and the team adjusted the design, avoiding what could have been a costly oversight. The simple act of noticing a teammate's silence – *listening beyond the explicit dialogue* – both solved a technical issue early and reinforced trust on the team. The junior engineer later said, "I really appreciated that my colleague noticed I was quieter than usual. It showed me that in our team, concerns don't have to be shouted to be heard." This kind of attentiveness sends a message that everyone's input is valued, boosting psychological safety.

In summary, active listening and keen observation allow an engineer to tune into the team's frequency. By being fully present in conversations and reading the nonverbal signals, you demonstrate respect. Colleagues and clients will sense that you genuinely care about what they have to say. This lays the groundwork for trust. As an active listener, you'll often catch the "little" issues (a teammate's confusion, a client's dissatisfaction) before they become big problems. It's a skill that might not come naturally to those of us used to focusing on machines more than people, but with practice it becomes an engineer's secret weapon for better collaboration and problem prevention.

Empathy in Action: Putting Yourself in Others' Shoes to Understand and Support

Listening is the first step; truly understanding and sharing the feelings of others is the next. That's where *empathy* comes in. Empathy is often defined as the ability to understand how another person feels and experiences the world – especially when their perspective is very different from your own. For engineers, who are trained to seek logical solutions, empathy might sound a bit abstract or "soft." However, empathy in action is extremely practical. It means putting yourself in your teammate's or end-user's position and seeing the situation through their eyes. This shift in perspective can be a game-changer in resolving conflicts, designing user-friendly systems, and offering support where it's needed most.

We often hear the phrase *"put yourself in their shoes."* But how exactly does an analytical mind do that? Consider this advice from an engineering blog on emotional intelligence: *"The simplest way of gaining a little perspective the next time an issue arises is to switch places with the other person and think about what's happening from their point of view. This way it's possible to understand each other enough to come to a resolution or offer advice."* In practice, this might involve pausing before reacting. For example, imagine a scenario where a project manager comes to an engineer with a last-minute change that upends some work. The engineer's gut reaction might be frustration or an urge to say, "No, that's not possible!" However, using empathy, the engineer could first consider the manager's perspective: perhaps the change is driven by urgent client feedback or a critical bug discovered in testing. By mentally "switching places," the engineer might realize the manager is under pressure too, not just trying to create chaos. Instead of snapping, the engineer can respond with understanding: "I realize this change is important. Let's figure out together how to accommodate it." This empathic approach turns a potential conflict into a problem-solving session.

Empathy isn't about being sentimental; it's a *practical skill* for teamwork. Research in organizational psychology has repeatedly found that empathy correlates with better performance and leadership effectiveness. In a global study of over 6,700 managers, the Center for Creative Leadership discovered that managers who exhibit empathy toward direct reports are seen as better performers by their own bosses. Why would that be? Because an empathetic leader (or teammate) listens to people's concerns, anticipates how decisions will impact others, and fosters loyalty and motivation. In engineering terms, empathy is like having a sensor for human factors – it alerts you to stress points in your team or user base that, if unattended, could fracture the system. One mechanical engineer described how learning about empathy changed his approach on the factory floor. If a technician came to him upset about a machine malfunction, his old response was to *immediately* dive into problem-solving mode, focusing only on the technical fix. Now, he first acknowledges the technician's frustration and asks questions to understand their experience: "That must have been frustrating when the conveyor stopped. What happened from your point of view?" He noted, "*It felt odd at first to talk about feelings at work, but it built trust. The technicians know I actually care, so they're more open about issues.*" By walking in their shoes momentarily, he not only fixes the machines but also addresses the human frustration, preventing resentment and reinforcing team morale.

Empathy is equally crucial when dealing with clients or end-users of engineering work. Engineers often design systems for people who don't share their technical background. Empathizing with end-users means understanding their needs, limitations, and emotions as they interact with a product or system. A classic case study comes from a software development team that built a complex data analysis tool. The tool worked perfectly from a technical standpoint, but adoption was abysmal – users were frustrated and intimidated by it. The breakthrough came when one developer spent a day sitting with a user (a non-technical analyst) and literally watched them use the tool, noting every point of confusion and frustration. By empathizing with the user's experience – feeling the confusion of too many options and the stress of potential errors – the developer realized that the interface was overwhelming. The team redesigned the user interface with simpler workflows and helpful prompts. This time, they even had someone role-play as a *new user* in testing, to ensure they truly saw it from a fresh perspective. The result was a product that users actually loved because it *fit their mental model.* As the developer put it, *"Instead of telling users to think like engineers, we learned to think like our users."* This is empathy in action:

taking on the user's perspective led to a far better solution.

Beyond usability, empathy can resolve interpersonal tensions on engineering teams. In high-pressure environments (think of a silicon chip design team close to a tape-out deadline, or a civil engineering crew rushing to finish a bridge before winter), stress runs high. Differences in working styles or communication can spark conflict. In such moments, an engineer skilled in empathy will pause and ask: *Why is my colleague acting this way? What might they be feeling right now?* Perhaps that terse code review comment from your teammate wasn't because they dislike you, but because they themselves are under immense time pressure or received bad news. One anonymous software engineer shared a quote about a conflict he defused with empathy: *"My teammate blew up at me over a small bug. I was taken aback, but then I remembered he'd been working nights on a big feature. I said, 'I know you've been working so hard and you're exhausted. I'm sorry about the bug – let's see how I can help fix it.' His attitude completely changed after that. We went from arguing to actually solving the issue together."* By validating his teammate's feelings (exhaustion and stress) instead of firing back, this engineer turned a blow-up into a bonding moment. The teammate later thanked him for understanding, saying he hadn't realized how stressed he appeared.

Empathy does not mean coddling poor performance or always agreeing with others. It simply means *understanding* the emotional context. With that understanding, an engineer can communicate more effectively. You might still have to give a colleague tough feedback or tell a client "no" to an unrealistic request – but if you do it with empathy, acknowledging their perspective, the message will be received far more positively than if you dismiss their feelings. In fact, empathy often opens up more options for solutions. By understanding *why* a coworker or client is upset, you can address the root cause rather than just the surface complaint.

In engineering teams, cultivating empathy has tangible benefits. Teams with empathetic members have been found to have lower conflict and increased cooperation. In one experiment on group problem-solving, teams that experienced positive emotional contagion (essentially, members sharing and mirroring empathetic, good-mood signals) ended up with *improved cooperation, less conflict, and higher perceived performance* on their tasks. Empathy, by helping team members feel understood, creates a more supportive atmosphere where people are willing to help each other. As an added bonus, empathy can *spark creativity* – when you see a problem through someone else's eyes, you might discover a new solution that you'd never have considered from your own viewpoint.

Engineers don't have to sacrifice logic for empathy; rather, empathy can inform our logic with better context. It's like getting additional input parameters for a problem. By integrating empathy into day-to-day interactions – asking questions, actively listening to responses, and imagining oneself in the other person's scenario – even the most analytical engineer can become adept at the human side of problem-solving. And as we've learned, that can be the difference between a project that *technically* meets requirements and one that genuinely satisfies and even delights the people involved.

Embracing Diversity in Thought: Valuing Different Backgrounds and Work Styles

Modern engineering is a team sport played on a global field. It's increasingly common for a project to involve a software developer in India, a UX designer in Germany, a data scientist in the US, and a project lead in Brazil, all working together. Even within a single office, teams are composed of individuals from diverse backgrounds – culturally, demographically, and in terms of professional training. Social awareness for an engineer thus also means cultural awareness and inclusivity: recognizing and respecting that people will think and communicate differently. By embracing this diversity in thought, an engineering team can unlock a wealth of creativity and ensure everyone feels valued.

Why is diversity so powerful for problem-solving? Imagine a team where everyone has virtually the same training and life experience – they might all approach a problem in the same way, potentially overlooking alternative solutions. Now imagine a team with a mix of backgrounds: say, a computer scientist, a cognitive psychologist, an artist, and an engineer who grew up in a developing country. The variety of perspectives can lead to more creative ideas and help the group *avoid blind spots.* Research backs this up: teams made up of diverse members make smarter and more accurate decisions on average than homogeneous teams. One engineering manager put it bluntly from experience, *"A diverse team yields better ideas, period."* When you bring together different viewpoints, each person might catch something others miss. For example, an algorithm development team at a tech firm was predominantly composed of young, single developers. They built an AI scheduling assistant with certain assumptions about work hours. It wasn't until a new team member, who was a working mother, joined that they realized the tool was annoyingly inflexible for people with childcare responsibilities. Her perspective led to new features (like smarter rescheduling and downtime between meetings) that made the product work for a much broader range of users. The team credited this improvement to having a voice that represented a different life experience.

Embracing diversity in thought also means fostering an inclusive environment where everyone feels heard. It's not enough to assemble a diverse team; engineers must cultivate a team culture that genuinely values each person's input. This ties closely to the concept of *psychological safety*. Psychological safety is the belief that you won't be punished or humiliated for speaking up with ideas, questions, or concerns. Google's extensive Project Aristotle study on team effectiveness found that psychological safety was the number one factor in distinguishing high-performing teams. In a team with high psychological safety, people feel safe to "be themselves" and voice divergent opinions. For an engineering team, that means a junior programmer can point out a potential flaw in the senior architect's design without fear, or an intern from a different country can share a novel idea even if it contradicts the conventional approach. When team members feel *included* and respected, they engage more deeply. One quote from a senior engineer captures this: *"In our team meetings, everyone's ideas get on the table. Even if an idea is a long shot, we don't shut it down immediately. Often that spark from left field inspires the real solution. Plus, people want to contribute when they know they'll be heard."* In short, inclusive teams leverage all their talent, not just the loudest or most senior voices.

Being culturally aware is a big part of this. Different cultures have different communication styles and values. A socially aware engineer on a global team will take the time to learn about colleagues' backgrounds. For example, in some cultures people may be less likely to speak bluntly or disagree with a proposal in a public meeting. Interpreting silence correctly is important – it might not mean consent; it could mean people are deferring to authority or feel uncomfortable speaking up. A team leader working with international colleagues noted that in his video conferences, his American and Western European teammates would readily argue their points, while some teammates from East Asian cultures were quieter. He realized they had excellent ideas, but they shared them privately or only when asked, due to a cultural emphasis on politeness and hierarchy. To embrace those differences, he started explicitly inviting input from quieter members in a respectful way, and sometimes using anonymous brainstorming tools where everyone could contribute without feeling put on the spot. The result was more balanced participation and a richer set of ideas. *"Once we created a space that fit everyone's style, the innovation just took off,"* he reported. The team went on to patent a solution that, by their own admission, *"none of us would have come up with alone."*

There's also evidence that diversity can prevent costly oversights. A poignant example involved an engineering design group that lacked diversity and nearly released a product with a major flaw. The team of developers had built a website for a broad consumer audience. Technically, it was excellent – fast, slick, and feature-rich. However, the group was fairly uniform in age and abilities, and they initially failed to notice that their color scheme and font choices made the site very hard to use for people with visual impairments or even in certain lighting conditions. Fortunately, before launch, a colleague with low vision from another department tested it and pointed out the accessibility issues. His *different perspective* (both literally and figuratively) helped the team realize they had optimized for a best-case scenario (perfect eyesight, large monitors, good lighting) and neglected a whole segment of users. The team quickly adjusted the design (adding features like high-contrast mode and larger text options), ultimately delivering a product that served all users better. They acknowledged that without that colleague's input, they would have alienated many potential customers and possibly faced compliance issues. This story highlights how having diverse team members (or reaching out for diverse feedback) can save you from "groupthink" errors. As one article on engineering teams noted, a homogeneous team will tend to "build to their own experiences," but a diverse team ensures you don't miss critical elements for others.

Embracing diversity also brings a less obvious but very real benefit: improved team morale and enjoyment. Working with people from different backgrounds can be enriching and fun. It introduces you to new ideas, foods, jokes, and holidays. It fosters mutual learning. In the words of one engineering leader, *"Perhaps the most compelling reason to have a diverse team is that it's simply more fun ... It prompts more creative discussions and problem-solving."* When team members feel they can bring their "whole self" to work and that their uniqueness is appreciated, they are more engaged and motivated. Conversely, if someone feels they have to hide who they are (maybe an engineer with a different personality or background feels pressure to conform to a narrow mold), it saps energy and creativity.

So how can an engineer actively promote an inclusive, diverse-minded culture? It can be as straightforward as inviting input from everyone and crediting people's ideas. It also means checking our own biases – for instance, not dismissing an idea just because it came from someone junior or from outside our domain of expertise. It may involve celebrating differences: rotate meeting times to accommodate various time zones, or have a "show and tell" at lunches where team members share something about their culture or interests. Even small acknowledgments matter, like learning to pronounce everyone's name correctly or being aware of important cultural festivals and not scheduling major deadlines on those dates if possible. These actions send a signal that diversity is valued, creating a safe space for all voices.

Finally, diversity of thought isn't only about demographic differences; it also means valuing different *working styles and personalities.* On an interdisciplinary team, you might have one engineer who likes to brainstorm out loud and another who prefers to quietly prototype a solution and present it. One person might favor tried-and-true methods, while another always pushes for cutting-edge tech. Instead of viewing these differences as conflicts, high-EQ engineers see them as complementary strengths. The cautious planner and the bold innovator can balance each other, if each appreciates the other's contribution. A case in point: an automotive design team we heard about had both "visionaries" who dreamed up radical new concepts and more "practical" engineers who worried about manufacturability and cost. Early on, these groups clashed – the visionaries felt bogged down by pragmatism, and the practical folks felt the ideas were unrealistic. The breakthrough came when they set ground rules that both types of input were needed: brainstorming sessions were separated from filtering sessions. In brainstorming, wild ideas were welcome without immediate "how would we build that" criticism. In later sessions, the practical engineers helped refine and implement the best ideas. By respecting these different work styles, the team ended up designing a car model that was both innovative *and* feasible, a result neither side could have achieved alone. The lesson is that diversity in approach can drive creativity, as long as there's mutual respect.

To sum up, embracing diversity in thought equips engineering teams with a broader toolkit to solve problems. It cultivates an environment where everyone feels they belong and can contribute. This not only leads to better technical outcomes (smarter decisions, more robust and inclusive designs) but also nurtures a positive team climate where innovation thrives. In a logical world, it turns out that including *all* voices is the logical choice for success.

Reading Team and Organizational Culture: Understanding the "Vibe" and Dynamics

Every team and company has its own personality – a set of unwritten rules, norms, and "vibes" that guide how people interact. Being socially aware as an engineer isn't limited to one-on-one interactions; it also means tuning into the *collective* mood and dynamics of your team and organization. In other words, you should develop a radar for group emotions and culture. This section is about learning to read the room when you walk into a meeting, sensing the undercurrents in your team, and navigating the broader organizational politics *ethically and effectively*. Just as a skilled sailor reads the winds and currents before setting sail, a skilled engineer-leader reads the emotional currents and power dynamics of their environment to choose the best approach.

Daniel Goleman (the psychologist who popularized emotional intelligence) describes *organizational awareness* as having the ability to "read a group's emotional currents and power relationships", including identifying key influencers and informal networks within an organization. In practical terms, this means paying attention to who has influence beyond their job title, how decisions actually get made, and what the "unspoken rules" of behavior are in your company. Every workplace has unspoken rules. For example, the official policy might be that work-life balance is valued, but the unwritten norm is that everyone stays until the boss leaves. Or a company might say "We encourage innovation," yet the culture frowns upon taking risks or challenging the status quo. A socially aware engineer picks up on these nuances. Knowing the real culture helps you strategize the best way to introduce ideas or make changes without stepping on landmines.

Consider reading the team vibe on a smaller scale. Suppose you join a new engineering team. In the first weeks, beyond learning the technical ropes, observe how the team operates. Is it a tense, heads-down atmosphere where people rarely joke or chat? Or is it lively and collaborative? Do decisions happen in big meetings or in quiet one-on-one conversations after the meeting? For instance, you might notice that on your team, whenever a new idea is raised in a meeting, everyone glances at a particular senior engineer before reacting. That tells you this senior engineer is an informal authority – getting their buy-in early would be wise. *Organizational awareness* in this case helped you identify an influencer. Goleman notes that people skilled in this competency can sense the informal networks and know "how to find the right person to make key decisions and how to form a coalition to get something done". This is the ethical side of "office politics" – it's not about gossip or manipulation; it's about understanding relationships and channels of communication so you can be effective in driving your ideas or projects. An engineer with great technical insight might still fail to get their proposal approved if they ignore the human network – for example, forgetting to consult the veteran technician whose support could sway the team. Social awareness teaches us that sometimes *how* an idea is introduced matters as much as the idea itself.

Another aspect is knowing when the team needs encouragement or a morale boost. Teams go through phases: there are crunch times, victories, setbacks, and even boredom during maintenance periods. A socially aware team member can gauge the emotional climate. Let's say the team has been working brutal overtime for two weeks to meet a deadline (a scenario all too familiar in engineering crunch times). You might sense weariness and low morale – people's tempers are shorter, humor is sparse, and folks look exhausted in the morning stand-up. Recognizing this, a savvy engineer might organize a small gesture, like bringing in coffee and donuts, or initiating a fun 10-minute break activity to lift spirits.

Perhaps you recall that the last push like this, someone organized a mini ping-pong tournament one evening, and it really cheered everyone up. Sensing the emotional low point allows you to respond helpfully. Even simply voicing appreciation can bolster the mood: "Hey everyone, I know it's been a tough week, but I have to say, the dedication I'm seeing is incredible. I'm proud to be on this team." Such words can energize a team if they're heartfelt and timely. In contrast, being tone-deaf to the team's vibe can alienate you. Imagine giving a long, abstract technical lecture to the team when they're stressed and behind schedule – it would likely frustrate them. Instead, noticing that stress, you might keep meetings short and focused, or defer non-urgent debates to later. Timing and tone are part of reading the culture.

Group dynamics are also about how people relate to each other in the team. Is there underlying tension between certain departments or individuals? Maybe the engineering team and the sales team have a history of friction (a common scenario, as one engineer quipped: "sales sells features we haven't built yet, and engineering gets annoyed"). Knowing this, if you're an engineer preparing to propose a change that will affect sales, you might first approach a friendly person in sales to get their perspective, rather than springing it on them in a big meeting. Or within a software team, maybe two senior developers have a rivalry that everyone knows about (except perhaps those two!). As a project lead, you might deliberately assign them tasks that don't put them in direct conflict, or you might have a candid talk to clear the air. Essentially, you are reading the interpersonal chemistry and taking steps to keep the team cohesive.

It's also important to understand your organization's wider culture. Is it a risk-averse culture where failures are hidden, or an open culture where people admit mistakes freely? If it's the former, as a socially aware engineer you might work on creating a micro-culture in your team that encourages learning from mistakes in a safe way (so that team members don't hide problems). If it's the latter, you know that you can be frank about setbacks and ask for help without losing face. Different companies also have different attitudes toward hierarchy. In a very hierarchical organization, ignoring protocol (like bypassing your boss to talk to their boss) might be seen as a serious breach. In a flat startup, that same action might be normal. One engineer recounted how, early in his career at a large traditional engineering firm, he emailed a suggestion directly to the division head – only to receive gentle feedback that this was not how things were done there; ideas should go through one's manager. He learned to navigate that culture by first convincing his manager to champion the idea. Years later, when he moved to a young tech startup, he initially hesitated to speak up in brainstorming sessions because he was used to deference; he then realized this new culture expected and valued direct input from everyone, regardless of title. The faster you can read the organizational norms, the more effectively you can operate within them (or influence them).

Another facet of organizational awareness is recognizing the emotional climate of the workplace. Emotions can spread in a group – a phenomenon psychologists call *emotional contagion.* If the overall vibe at the company is anxious (say there are rumors of budget cuts or a round of harsh product reviews), that mood will seep into team interactions. A socially aware person will pick up on this ("The usual buzz in the office is quieter these days; people seem on edge") and can respond appropriately. For example, if you sense general anxiety, it might be a good time for a team huddle to acknowledge challenges and reinforce solidarity: "Yes, the last demo failed, but we're in this together and we've solved tough problems before." Research has shown that positive emotional contagion – in plain terms, shared upbeat mood – can lead to improved cooperation and even better performance among team members. Effective leaders consciously try to set a positive emotional tone. On the flip side, if morale is dangerously low (perhaps after a project cancellation or during crunch burnout), a wise engineer will know that pushing people harder could backfire. Sometimes the team needs a breather or some acknowledgment of their feelings. Simply *sensing* that and empathizing at the group level ("I know everyone is discouraged right now; it's okay to feel that way – maybe we can do something fun Friday to recharge") can prevent disengagement.

It's worth noting that "reading the vibe" is not the same as pandering to it. It doesn't mean you avoid all conflict or never introduce bold ideas. It means you introduce them in the right way and at the right time. Suppose you have a revolutionary idea for re-architecting a product – it could be hugely beneficial, but it's also going to upset the status quo. If you propose it on a day when the team is already overwhelmed firefighting issues, it might be dismissed outright. But if you bide your time until after a successful release when people are more relaxed and receptive, you might get a very different reception. Or, perhaps you sense that *you* alone pushing this idea might trigger resistance (the "not invented here" syndrome). Using organizational awareness, you identify a respected ally in the team who also sees the need for change, and you enlist them to co-present the idea. By forming a small coalition, you increase the idea's chances. These are strategic moves that take into account the human landscape.

Navigating office politics ethically essentially boils down to leveraging your social awareness to build alliances and communicate effectively, without deceiving or manipulating. It's the difference between a scheming approach ("I'll play people against each other to get my way") and an emotionally intelligent approach ("I understand everyone's interests and I'll find a win-win by getting the right people on board"). Colleagues generally appreciate the latter, even if they don't articulate it, because it respects the *unwritten rules* and relationships that hold the organization together.

In conclusion, being able to read your team and organization is a powerful skill for an engineer. It enables you to sense when to lead and when to listen, when the team needs a push and when it needs a pat on the back. It helps you figure out the best way to introduce new ideas and avoid stepping on toes unnecessarily. Engineers operating in isolation can get blindsided by cultural resistance, but engineers high in organizational awareness navigate those currents like seasoned sailors. They become the people who just *seem to know* the right timing for everything – when to ask for resources (maybe after the company announces a big new client), when to raise a concern (perhaps privately with a manager rather than in public, if that fits the culture), and how to adapt their style to different groups. This skill keeps teams cohesive and motivated. It's the glue between individual empathy and broad leadership.

To wrap up Chapter 4, remember that social awareness bridges the logical and the human. As engineers, when we listen actively, practice empathy, embrace the wealth of diversity around us, and read our environment astutely, we are not abandoning logic – we are applying it to the complex system that is human teamwork. These skills allow us to preempt problems, create trust, spark innovation, and guide our teams through the turbulence of projects and organizational change. In the logical world of engineering, emotional intelligence is not a foreign element; it's the catalyst that makes the whole solution work. Each story and example here, from listening beyond words to feeling the pulse of your company, shows that mastering these "people skills" can elevate an engineer from a good individual contributor to a great team player and leader. And that is ultimately what engineering, at its highest level, is all about: collaborating with others to build something greater than any one of us could build alone.

Chapter 5

Relationship Management – Communication and Collaboration

Engineers often pride themselves on logical problem-solving and technical expertise, but engineering success isn't **just** about writing great code or designing robust systems. It's also about how well you work with other people. This is where emotional intelligence in *relationship management* comes in – the ability to use awareness of your own and others' emotions to guide interactions toward positive outcomes. In a field driven by logic, engineers who master these "people skills" stand out. They lead projects that don't just work technically, but also run smoothly because team members feel understood, motivated, and united. In this chapter, we explore how high-EQ engineers communicate clearly, exchange feedback constructively, resolve conflicts calmly, and build collaborative relationships that amplify everyone's success.

Clear Communication for Technicians

One hallmark of emotionally intelligent engineers is clear and adaptive communication. As technical professionals, we swim in complexity every day. But the true test of understanding a complex idea is being able to explain it in simple terms. Whether you're detailing a software architecture to fellow developers or summarizing a project update for a non-technical executive, clarity is key. Jargon and acronyms might impress peers, but they can alienate clients or stakeholders. High-EQ engineers know how to "translate" technical concepts for different audiences without sounding condescending. They tailor their language and level of detail to what others need to know.

As a popular saying goes, *"If you can't explain it simply, you don't understand it well enough."* This quote (often attributed to Albert Einstein) resonates in engineering: breaking an idea down forces you to grasp its essence, and it ensures your audience grasps it too.

Use analogies and stories: A great strategy for clear communication is using analogies from everyday life. For example, one engineer described a distributed system's components interacting as "like a relay race team handing off a baton" – a simple image that helped a non-technical manager immediately visualize how the parts worked together. Emotionally intelligent communicators also watch their listeners' body language and questions for feedback. If a client's eyes glaze over when you mention "API latency optimization," that's a cue to pause and rephrase. Perhaps you compare it to traffic flow, saying something like: "Imagine fewer cars on the road means faster travel; similarly, fewer data requests make our application respond quicker." By relating new concepts to familiar ideas, you bridge the understanding gap with empathy for your audience's perspective.

Avoid the curse of knowledge: It's easy for experts to forget what it's like *not* to know something – a cognitive bias known as the *curse of knowledge*. Engineers high in EQ consciously step into the shoes of someone without their background. They avoid unnecessary technical detail and define terms that might be new to others. For instance, instead of saying "We need to refactor the monolith into microservices for better scalability," you might add: "...which means breaking one big application into smaller independent pieces that can be developed and scaled individually." A little extra explanation can prevent big misunderstandings.

The costs of poor communication can be huge. A classic cautionary tale comes from NASA: in 1999, the Mars Climate Orbiter probe was lost because one engineering team used imperial units (like feet and pounds) while another used metric units, and nobody caught the discrepancy. In a sense, the spacecraft was "lost in translation". This $125-million mistake underscores why clear, consistent communication matters even among technical teams. When engineers fail to make sure everyone is "speaking the same language" – literally or figuratively – small misunderstandings can snowball into costly problems.

On a day-to-day level, clear communication builds trust and efficiency. When you explain the *why* behind your solution in plain language, teammates and stakeholders are more likely to buy into the vision. They won't have to nod along pretending to understand; they'll actually get it, ask better questions, and offer useful input. By being articulate in writing (think well-structured design docs or emails) and in speech (from stand-up meetings to client demos), you ensure everyone is aligned. In short, mastering clear communication turns an engineer from a lone technical genius into a powerful collaborator. It's a skill that makes the brilliant ideas in your head accessible and actionable to the people around you.

Constructive Feedback and Dialogue

Another key aspect of relationship management is how we give and receive feedback. In the engineering world, this often plays out during code reviews, design critiques, or post-mortems. With low emotional intelligence, these interactions can turn into defensive battles or hurt feelings. High-EQ engineers, however, use feedback as a tool for growth rather than a weapon. They deliver criticism in a positive, respectful manner and are just as open to receiving critique from others.

Giving feedback without offense: The golden rule is to critique the work, not the person. Imagine you're reviewing a colleague's code. Saying *"You wrote this function inefficiently"* can make them feel personally attacked. Instead, a constructive phrasing might be, *"This function could be more efficient – perhaps we can refactor it to reduce the number of database calls."* Now the focus is on the code itself and an invitation to improve it together, rather than on blame. Many experienced reviewers also phrase suggestions as questions: *"What do you think about caching this value?"* or *"Have you considered this approach?"* This opens up a dialogue and is less confrontational, showing respect for the author's thinking.

Tone and wording make a huge difference. High-EQ engineers avoid loaded words such as "obviously" or "simply" in their comments, since these can come across as condescending. They know that words like "just" or "wrong" can trigger defensiveness. They also steer clear of sarcasm in written communication, recognizing that without vocal tone, sarcasm easily misfires and hurts feelings. By being mindful of language, emotionally intelligent feedback-givers ensure their message is heard as helpful, not as an attack.

Turning feedback into growth: Emotional intelligence also shows in how we receive critique. It's natural to feel a twinge of defensiveness when someone points out flaws in our work – after all, we pour a lot of ourselves into that code or design. But high-EQ individuals practice pausing that emotional reaction and listening for the useful insight in the feedback. Instead of reacting, they reflect. For example, if a teammate says your server module isn't scaling well, an EQ-driven response might be: *"Thank you for catching that. Do you have thoughts on possible fixes? Let's dig into it."* This approach treats feedback as a gift aimed at improving the project, not as a personal slight.

Cultivating this mindset can transform team culture. When engineers start viewing feedback in code reviews as shared problem-solving rather than as grades on a report card, everyone wins. Teams that emphasize constructive, respectful feedback tend to continuously improve and innovate. They create what one open-source community famously called a "culture of review" where each critique is an opportunity to learn.

In fact, the importance of respectful dialogue was highlighted by a well-known case in the tech world: the leader of a major open-source software project (known for his blunt communication style) publicly apologized after realizing his harsh feedback was driving contributors away. He helped adopt a new code of conduct that encouraged "accepting constructive criticism gracefully" and "being respectful of differing viewpoints". It was a wake-up call that even brilliant engineers achieve more when they communicate with empathy and respect.

Pragmatically, giving feedback with emotional intelligence means you're not just pointing out problems – you're motivating better solutions. Likewise, receiving feedback with grace means you'll get better over time and people will *want* to work with you. A code review comment like *"Good thought here. Maybe we can simplify by doing X... what do you think?"* can spark a learning conversation rather than a confrontation. By making feedback a two-way dialogue, engineers turn what could be tense moments into trust-building exercises. Over time, this fosters a team ethos where feedback isn't feared but valued – a driver of continuous improvement and mutual respect.

Conflict Resolution Skills

No matter how great the communication or how positive the feedback culture, disagreements will arise in any team. Engineers might clash over the best design approach, or personal frictions might develop under deadline stress. What sets high-EQ professionals apart is how they handle these conflicts. Instead of letting disagreements fester or explode, they navigate them calmly and constructively, often turning conflict into an opportunity to reach a better outcome.

Consider a scenario: two engineers on a team strongly disagree on which database to use for a new product. The discussion in a meeting starts getting heated. A high-EQ individual in this situation will notice their own rising frustration and take a mental step back before responding. This moment of self-regulation (a skill covered in earlier chapters) prevents the conflict from becoming personal. Instead of snapping *"You just don't get it!"*, an emotionally intelligent engineer might say in an even tone, *"I see we have different approaches. Let's break down the core goals here and figure out what we both want."* By managing their own emotions, they set a tone that encourages calm, rational discussion.

Listen to understand, not to win: Conflict resolution heavily depends on listening – really listening. Often in technical disputes, each side is so busy defending their solution that they stop hearing the other's points. Stephen Covey famously noted, *"Most people do not listen with the intent to understand; they listen with the intent to reply."*. High-EQ engineers strive to avoid this trap. They ask open-ended questions like "Can you walk me through why this approach meets our requirements?" and paraphrase the response: "So, you're concerned the API response time might be too slow with that method, is that right?" Paraphrasing ensures you truly get the other person's perspective and makes them feel heard. Often, just being heard diffuses a lot of tension. This active listening also uncovers that many disagreements aren't about *goals* – which are usually shared – but about differing assumptions or methods. Once each side understands the other's reasoning, the team can focus on what they agree on and where the true points of contention lie.

Find common ground and focus on solutions: Emotionally intelligent conflict resolution means guiding the conversation toward solutions, not blame. Instead of fixating on "who's right," high-EQ collaborators look for the *common ground*. In the database debate above, the common goal might be: "We all want a reliable, scalable system that meets our users' needs." Acknowledging that out loud ("We're on the same team here") creates a collaborative atmosphere. From there, it's easier to evaluate options objectively – perhaps by listing pros and cons together, or agreeing to test each approach on a small scale. Sometimes a creative compromise emerges: for example, using one database as the primary store and the other for a specific feature, combining the strengths of both proposals. By keeping discussions respectful and solution-focused, conflicts can actually spur innovation. Two clashing ideas might merge into an even better one when people are open-minded. In contrast, low-EQ handling – like personal attacks or stubborn stonewalling – breeds resentment and stifles creativity.

There are also simple frameworks engineers can use for navigating friction. One is to separate impact from intent. If someone's action upset you, explain the impact it had on you instead of immediately accusing their intent. For example: *"When the deployment failed and I found out from the client before the team, I felt blindsided. I know you likely didn't intend that, but can we improve our communication process so it doesn't happen again?"* This kind of message focuses on the issue and its effect, rather than labeling anyone as the villain. It opens the door to a constructive conversation about process, instead of a defensive argument about personal fault. Along similar lines, remember to attack the *problem*, not the person. In disagreements over ideas, avoid ad hominem remarks (e.g. "Only an idiot would design it that way"). Stick to the facts of the technical issue.

High-EQ teams sometimes even establish ground rules for handling conflicts. For instance, they might agree that if a debate gets too heated, they'll take a short break and resume with a cooler head, or they'll invite a neutral third party (like a team lead or architect) to mediate complex issues. The goal is always to resolve the issue while preserving relationships. When conflict is handled with empathy and respect, teammates learn that they can disagree without fear of reprisal. In fact, having the confidence that conflicts will be managed fairly is a big part of what creates *psychological safety* on a team. It allows people to surface concerns and challenge ideas, leading to better technical decisions in the long run.

Collaboration and Influence

The ultimate payoff of strong relationship management is seen in collaboration – how well you work with others toward a common goal – and in your ability to positively influence those around you. In engineering, no matter how brilliant you are individually, significant accomplishments are almost always team efforts. High-EQ engineers recognize this and actively cultivate trust, empathy, and reliability, which in turn make them excellent collaborators and informal leaders.

Building trust through empathy and reliability: Trust is the currency of effective collaboration. Team members need to know they can count on each other – not just to do competent work, but to have each other's backs. One famous internal study at Google, called *Project Aristotle*, found that the highest-performing teams weren't defined by the smartest individuals or the best technical skills, but by psychological safety – a shared confidence that the team is a safe environment for interpersonal risk-taking. In other words, when people feel safe to express ideas or admit mistakes without fear of ridicule, teams thrive. Trust and respect are at the core of this safe environment. For an engineer, demonstrating empathy is a powerful way to build trust. Something as simple as offering help when a teammate is struggling with a tricky bug, or actively soliciting input from a quiet team member, shows that you care about your colleagues' success. Over time, these small acts establish you as someone others want to work with.

Reliability is equally critical. Emotional intelligence helps here by making you aware of how your commitments and follow-through affect others. If you consistently deliver what you promise and communicate early when you can't, you establish a reputation for dependability. As a result, even without a lofty title, you become influential because people trust your word and judgment. One veteran engineering leader noted that *influence without formal authority* is built by earning trust through consistent technical excellence, honest communication, and reliable follow-through. When your teammates trust that you know your stuff and that you treat them fairly, you've earned a form of leadership by respect.

Sharing credit and enabling others: Collaboration is not a zero-sum game; in fact, the best collaborators actively make those around them better. High-EQ engineers take joy in the team's achievements and readily share credit. If a project milestone is reached or a difficult problem gets solved, they make sure to highlight everyone's contributions. This isn't just being "nice" – it has a practical effect too: openly sharing credit shows you're secure and supportive, which builds trust within the team. Colleagues feel seen and appreciated, which strengthens the relationship and motivates them to collaborate again. By contrast, if someone habitually hogs credit for successes, trust quickly evaporates and others become reluctant to work with them. For example, a senior developer might point out in a meeting, "We couldn't have fixed that outage so quickly without Ali's debugging script; it was crucial" – a simple acknowledgment that boosts morale and underscores that collaboration is valued.

Another aspect of being a great teammate is helping colleagues grow. Mentoring a junior developer, sharing knowledge, or pitching in to pair-program on a tough issue all show that you're invested in others' success, not just your own. High-EQ individuals are not threatened by others shining; they actually facilitate it. This generosity with time and knowledge creates an atmosphere of continuous learning and support. When a team member knows you genuinely want to help them succeed, they in turn trust your intentions and often go the extra mile for you when you need it. Over time, this supportive environment means information flows freely and problems are solved faster – everyone benefits. Plus, when you uplift others, they are more likely to support your ideas and initiatives, creating a virtuous cycle of mutual influence.

Influence without authority: One remarkable effect of being a strong collaborator with high EQ is that you can lead from any seat. Influence in an engineering organization doesn't only come from a job title; it comes from trust and respect. Consider a case of an early-career engineer who had an idea to streamline the build process at her company. She had no managerial authority, but she had built strong relationships across the development and DevOps teams by working empathetically and reliably with them on prior projects. When she pitched her idea, people listened. She incorporated others' feedback, addressed concerns collaboratively, and volunteered to prototype the solution. She soon rallied a small cross-functional group to adopt the new build tool. The initiative succeeded largely thanks to the trust she had earned – she essentially *led* that change through influence alone.

The formula often boils down to expertise + empathy = influence. You absolutely need solid technical credibility – people won't rally behind an idea that doesn't make sense – but emotional intelligence is what lets you communicate that idea in a compelling way and bring others on board. By listening to stakeholders' concerns, aligning your proposal with things they care about, and showing respect for every contributor's perspective, you make it *their* idea too. People support what they help create, so a high-EQ influencer involves others early and gives them ownership in the process.

In practical terms, if you want to increase your influence, focus on building relationships before you need them. That might mean taking the time to get to know engineers in other teams, understanding their challenges, and offering help or insights without any immediate agenda. Also, demonstrate integrity – if a production incident occurs on your watch, take responsibility openly rather than hiding the issue. Each of these actions is like a small deposit in a "trust bank account." Over time, as you accumulate trust and goodwill, you'll find that when you have a proposal or need to rally support, people are already inclined to hear you out and join forces.

Ultimately, engineers who excel in relationship management create a positive feedback loop in their careers. Their communication and collaboration skills make projects more successful (and more enjoyable to work on), which builds their reputation – and that reputation in turn gives them even greater ability to influence and drive future successes. By mastering clear communication, constructive dialogue, conflict resolution, and empathetic collaboration, you position yourself not just as a highly skilled engineer, but as a true team leader in a logical world that increasingly values the human touch.

Chapter 6

Leading with Emotional Intelligence – Guiding Teams with Heart and Logic

Leading an engineering team isn't just about hitting deadlines or optimizing code – it's also about understanding people. In a field driven by logic and data, the idea of bringing "heart" into leadership might seem counterintuitive. Yet experience and research show that emotional intelligence (EQ) is a critical factor that separates good leaders from truly great ones. When engineering leaders balance analytical thinking with empathy and self-awareness, their teams tend to thrive. Morale rises, trust deepens, and performance improves. Research shows that employees with empathetic, supportive managers are happier, more engaged, and less likely to burn out. By contrast, even brilliant technical minds can struggle as leaders if they lack emotional intelligence – a low-EQ boss might hit short-term targets by pushing people, but it comes at the cost of burnout and turnover.

This chapter explores how to guide teams with both heart and logic. We'll delve into what empathy-driven leadership looks like in a technical environment and how understanding your team's perspective can boost morale and loyalty. We'll discuss why the best engineering managers act as coaches and mentors rather than micromanagers, developing their people through support and feedback. Next, we'll look at building a culture of trust and innovation, where team members feel safe to speak up and take risks – an environment that fosters collaboration and creativity. Finally, we'll cover strategies for handling change and crisis with EQ, showing how honest, empathetic communication and composure during tough times keep a team's trust and focus. Throughout, the tone is conversational and professional, sharing stories and strategies to inspire you to lead with both your heart and your head.

Empathy-Driven Leadership: Using Empathy as a Powerful Leadership Tool

Leading with empathy means seeing beyond the code and recognizing the human beings behind it. For an engineer-turned-manager, it can be tempting to focus only on logical processes and assume everyone will respond purely to facts and figures. However, an empathy-driven leadership approach acknowledges that each team member has unique perspectives, emotions, and motivations. Using empathy as a leadership tool begins with actively listening to your team. Instead of just talking about project requirements or bug counts, an emotionally intelligent leader takes the time to ask, "How are you doing with this workload?" or "Is anything hindering your progress that I might not be aware of?" By opening up these conversations, you show your team that you truly care about their experience and well-being, not just the output of their work.

Empathy-driven leadership has tangible benefits – when team members feel understood and valued, their morale and loyalty skyrocket. They know their manager "has their back," which makes them more willing to go the extra mile. Numerous studies back this up – leaders with high emotional intelligence can significantly boost employees' job satisfaction and even on-the-job performance. Put simply, people perform better when they feel happier and supported at work. Conversely, a lack of empathy in leadership can be detrimental. Employees under managers who are cold or dismissive often feel disengaged, stressed, and ready to leave for better opportunities. As one business writer put it bluntly, people won't follow a leader who doesn't seem to care. In the engineering world – where talent is in high demand – losing good people because a manager failed to show basic understanding is an expensive mistake.

So how can you practice empathy in day-to-day leadership? It starts with small habits. Active listening is one: when a team member has a concern or idea, give them your full attention and really try to understand their point of view before jumping into problem-solving. Also, put yourself in their shoes. If a junior developer is struggling with a tough task, remember how it felt and offer the kind of help you wish you'd had at that stage. Empathy also means being aware of your team as *whole people*, not just as "resources." Notice if someone has been working overtime for weeks – they might be nearing burnout – or if a usually upbeat colleague seems down. Acknowledge these situations and adjust when possible – encourage a burnt-out teammate to take a day off, or lighten someone's load during a personal crisis. By taking care of your people in these ways, you send a powerful message that they are valued as human beings, not just interchangeable cogs.

Empathy-driven leadership builds a reservoir of goodwill and trust that pays off in the long run. Team members who feel their manager genuinely cares are more likely to stick around through tough times, be open about problems early, and trust the decisions being made. Empathy also fosters loyalty – employees remain committed to leaders who treated them with understanding, even when challenges arose. And loyalty translates to discretionary effort: the team that feels appreciated will put in that extra bit of creativity and energy when it's really needed. Using empathy as a leadership tool creates a positive feedback loop – high morale and loyalty lead to better performance and innovation, which in turn reinforce everyone's sense of trust and camaraderie.

Some may worry that being empathetic means being "soft" or less focused on facts. In reality, empathy is not a weakness – it complements logic. You aren't replacing data with feelings; you're integrating both to make better decisions. By understanding your team's needs and perspectives, you can implement solutions that actually work better because they account for the human element. A leader can care about people and still be rational about business decisions – in fact, that balance of heart and head is what inspires the most respect.

Coaching and Mentoring: Developing Your People with High EQ Leadership

One hallmark of an emotionally intelligent leader is the commitment to growing and developing their team members. In engineering, this means shifting from being the expert with all the answers to being a coach and mentor who empowers others to find answers themselves. Technical leaders with high EQ recognize that their role is not to micromanage every design decision, but to cultivate an environment where each engineer can learn, contribute, and shine. This coaching mindset requires skills like active listening, patience, and the ability to tailor your approach to individual motivations – all rooted in emotional intelligence.

Imagine a junior engineer writes some code that isn't up to standard. A low-EQ manager might swoop in, rewrite it themselves, and scold the engineer for the mistakes – leaving that person discouraged and hesitant to take initiative next time. An emotionally intelligent leader, on the other hand, treats it as a coaching moment. They meet with the engineer one-on-one, listen to their thought process, and then gently guide them toward a better solution (perhaps saying, "I see what you were going for; let's try a different approach together"). This turns a failure into a learning experience. The engineer leaves feeling supported rather than shamed, and they'll be more confident tackling the next challenge knowing their manager has their back.

A coaching-focused leader also gives feedback in a way that motivates rather than deflates. They make feedback timely, specific, and delivered with empathy.

Instead of a vague pat on the back or a harsh "fix this" critique, they explain what was done well and then suggest improvements in a constructive manner. They also deliver criticism in private rather than embarrassing someone in front of others, which preserves the person's dignity. When employees sense that feedback comes from a place of genuine care for their development, they become much more receptive to it.

Emotionally intelligent leaders additionally act as mentors, focusing on their people's long-term growth. They take time to discuss each team member's career goals, strengths, and aspirations. This shows that as a leader, you care about their future, not just their output today. If you see potential in someone – say they have great communication skills or leadership instincts – you encourage it and help them find opportunities to develop those talents. When people see that you're genuinely invested in their success, it builds tremendous goodwill and loyalty.

Crucially, emotionally intelligent leaders avoid micromanaging. Hovering over every detail or second-guessing each decision signals a lack of trust and can smother motivation. Instead, high-EQ leaders delegate meaningful responsibilities and then step back, remaining available as a safety net if needed. If something goes off track, they don't swoop in with blame – they offer support and guidance to help the person course-correct. This shows the team that their leader trusts them and turns problem-solving into a collaborative effort rather than a top-down directive. Over time, this approach builds a more confident, self-motivated team – people put in their best effort and tend to stick around when they feel trusted and supported.

Building a Culture of Trust and Innovation: Fostering a Safe and Collaborative Environment

A team that trusts each other and their leadership is a team that can truly innovate. Psychological safety means team members feel safe to speak up with ideas or admit mistakes without fear of punishment. When leaders foster this kind of safety, people are far more willing to share their creativity and point out problems without looking over their shoulder. In fact, Google's Project Aristotle found that psychological safety was the number one predictor of high-performing teams. In other words, you can have a group of brilliant engineers, but if they're afraid to raise concerns or suggest bold ideas, the team's performance will suffer. Conversely, even a relatively junior team can outperform expectations if everyone feels free to contribute and knows they won't be ridiculed or penalized for speaking up.

Building this kind of trust begins with leadership. As the saying goes, the boss "makes the weather" for the team. If you model openness and respect – for example, explaining the reasons behind decisions and truly listening to feedback – those behaviors will spread. Team members see that you value their input and treat them fairly, which encourages them to speak up. On the other hand, if a leader dismisses feedback or reacts angrily to bad news, people will conclude it's safer to stay silent. Over time, openness from the top creates a culture where information flows freely and issues come to light before they become big problems.

Encourage open, honest dialogue within the team. You might set up a regular forum where everyone is invited to share concerns and ideas – and crucially, respond to mistakes or bad news with a learning mindset instead of blame. This shows that errors are opportunities to improve, not something to hide. Over time, teams with high trust catch and solve problems faster because people speak up early rather than staying silent.

Leading by example is also crucial. If a leader is always emailing at midnight or panicking at every setback, the team will feel expected to do the same – leading to burnout and anxiety. In contrast, a leader who sets healthy boundaries and stays calm under pressure sends the opposite signal. Team members take their cues from the boss; as the saying goes, the leader often "makes the weather" for the team. By handling stress with composure and modeling a sustainable work style, you foster an atmosphere of balance, stability, and trust.

A high-trust culture reaps huge rewards. When people feel safe and valued, they collaborate more freely and are willing to propose creative ideas – making the team more innovative and resilient. They are also more likely to stick around, since few will jump ship from a workplace where they feel respected and inspired. In the end, while technical skills are vital, it's often the emotional intelligence of leaders that turns a merely efficient engineering team into a truly exceptional one.

Handling Change and Crisis with EQ: Leading Through Uncertainty with Heart and Logic

Change and crisis are the moments that test any leader's mettle – and this is where emotional intelligence truly shines. Whether it's a major organizational change, a painful event like layoffs, or an unexpected crisis such as a critical system failure, an engineering leader must guide their team through turbulence with both heart and logic. The goal is to maintain trust and stability even when circumstances are tough. To do that, honesty, empathy, and composure are your greatest allies.

In any major change or crisis, communicate early and truthfully – if people are left in the dark, their anxiety will only grow. An emotionally intelligent leader knows that trust is built on transparency, even when the news is bad. Your team would much rather hear difficult news directly from you than through rumors. If you don't have all the answers yet, it's okay to admit that – tell them what you do know and promise to update them when you learn more. By being forthright, you show respect for your team's need for honesty, which helps maintain trust even in tough times.

Equally important is to lead with empathy throughout the turmoil. Before jumping into problem-solving mode, take time to acknowledge and listen to people's concerns. Let team members express their frustrations or fears, and show that you truly hear them. A simple validation like saying "I know this is tough, and it's okay to feel upset" can defuse a lot of tension. For example, if you must ask the team for a big extra effort, acknowledge the sacrifice and thank them sincerely. These human touches go a long way toward keeping morale intact when things are hard.

Maintain your composure and self-control. Your team looks to you as an anchor, so if you panic or lose your cool, they will too. Use your self-awareness to manage stress – take a moment to breathe or step aside to collect yourself when needed. Remember that keeping yourself in good shape – with rest and steady nerves – is necessary so you can support your team. When your team sees you staying steady and resilient, it reassures them that they can get through the challenge as well.

Even in dreaded scenarios like layoffs, an emotionally intelligent leader will handle the situation as humanely and honestly as possible. They deliver the news with empathy and transparency – ideally in person or in a one-on-one call – and they acknowledge the pain and uncertainty people feel. By treating those affected with respect and offering support to those who remain, such a leader preserves trust even amid bad news. In contrast, a cold, impersonal approach (for example, employees finding out they're laid off only by suddenly losing system access) shatters morale and trust. Leading with heart during these times doesn't erase the difficulty, but it keeps a foundation of respect and loyalty that will help the team recover.

Emotional intelligence truly transforms the way engineers lead. By combining heart and logic, you create a work environment where people feel motivated, supported, and free to do their best work. Empathy-driven leadership boosts morale and loyalty. A coaching mindset develops a stronger, more capable team. A culture of trust sparks innovation. And steady, compassionate guidance in crises preserves unity and trust. In a logical world that often prizes technical prowess above all, EQ is the "soft" skill that delivers hard results. Leading with emotional intelligence means leading with both the mind and the heart – and that balance is the key to guiding teams to success.

Chapter 7

Lifelong Growth – EQ in Your Career and Life

Emotional intelligence isn't a one-and-done skill – it's a lifelong journey of growth. Just as technology evolves and engineers diligently update their technical skills, EQ (Emotional Quotient) requires continuous learning and adaptability over the course of your career and life. In this chapter, we explore how treating EQ as an ongoing development makes you a more effective engineer and a more fulfilled person. We'll look at how you can keep refining your emotional savvy from your first job to the executive suite, how EQ skills enrich your personal life, why blending IQ and EQ yields the best results, and how emotional intelligence will future-proof your career in a changing, digital world. The tone here is conversational and motivational – because mastering EQ is an inspiring quest that can elevate every aspect of your life.

Continuous Learning and Adaptability

Emotional intelligence is not a static trait you either have or don't – it's a set of skills you can continually develop. Think of it like learning a new programming language or framework: there's always another level of proficiency to reach. Successful engineers treat EQ as a lifelong development journey. In practice, this means actively seeking opportunities to learn about yourself and others, and being open to growth in your interpersonal skills. One engineering manager described emotional intelligence as *"a journey, not a destination"* – you never truly "finish" learning it. In fact, leadership experts emphasize that EQ requires ongoing effort throughout one's life, with deliberate practice and feedback along the way. Embracing this continuous-learning mindset for EQ will make you as adaptable in dealing with people as you are in dealing with technology.

Adopt a growth mindset toward EQ. A growth mindset, a term popularized by psychologist Carol Dweck, means believing your abilities can be developed through dedication and hard work. Apply this to emotional intelligence: view every project, team, and even conflict as a chance to get better at communication and empathy. One tech CEO famously pushed his company to shift from a "know-it-all" mentality to a "learn-it-all" culture – in other words, to value curiosity and learning over brash certainty. He echoed Dweck's philosophy by noting that *the person who lacks some natural ability but keeps learning will eventually outshine the so-called genius who stops growing.* The same applies to EQ: even if social skills don't come naturally at first, a commitment to keep improving them will pay off. Emotional intelligence supports this growth mindset by helping engineers stay flexible and creative amid new challenges. With strong self-awareness, for example, you can quickly spot areas to improve (maybe you notice you get defensive during code reviews) and then actively work on them. Over time, these incremental improvements compound into major gains in your interpersonal effectiveness.

Seek mentors, coaches, and role models for EQ development. Just as you might have a senior engineer guiding you in mastering a new coding technique, it helps to have mentors for the "people" side of engineering. Early in your career, you could seek out a more experienced colleague who exemplifies great teamwork or a manager known for empathy, and learn from their behavior. Don't be afraid to ask for feedback on your communication and leadership style – it's the equivalent of debugging your emotional skillset. Many top performers invest in coaching or training programs to boost their EQ, recognizing that these skills can be learned with practice and guidance. For instance, if you struggle with giving presentations, a coach might help you develop emotional self-regulation techniques to manage anxiety and convey confidence. By treating these soft skills as learnable, you empower yourself to grow. Remember, even leaders at the highest levels continue to refine their EQ. It's common to hear executives talk about learning from failures or tough conversations – they are effectively expanding their emotional intelligence through real-world experience.

Reflect on experiences and adapt as you advance in your career. The EQ skills you need most will evolve as you move from an entry-level engineer to a team leader to, perhaps, an executive. At the start of your career, active listening and teamwork might be your focus – for example, learning to take constructive criticism without defensiveness, or communicating your ideas clearly to colleagues from different disciplines. As you progress to leading projects, you might notice new emotional challenges: motivating a demoralized team, negotiating deadlines with empathy, or resolving conflicts between team members. Each stage offers lessons if you pause to reflect. After a big project, ask yourself: *How did I handle the stress? Did I communicate effectively under pressure? Where could I have been more understanding of others' perspectives?* By reflecting, you can identify what to adjust. Maybe you realize you were micromanaging out of anxiety – that's a cue to develop greater trust and delegation (an EQ skill for managers). Or perhaps you handled a crisis calmly – that's a strength to continue leveraging. Engineers who treat each experience, whether success or setback, as feedback for their EQ development end up far ahead of those who stick to "this is just how I am." In one real-world example, a software developer noticed that whenever a project slipped behind, he instinctively withdrew and avoided communicating the bad news. This led to confusion and mistrust on his team. After some mentorship and honest self-reflection, he worked on being more transparent and emotionally resilient. The next time challenges arose, he proactively discussed them with the team, maintaining trust – a clear improvement in his leadership effectiveness.

Continuous adaptability is key. The engineering world changes rapidly – new technologies, new team structures, even entire new industries emerge in a span of a few years. Technical skills you mastered five years ago might become obsolete, but your emotional intelligence only grows more valuable with time. By committing to continuous EQ learning, you ensure that you can *adapt* to whatever your career throws at you. For instance, if you switch from a technical role to a client-facing role, your empathy and communication skills will help you adjust quickly to understanding client needs. Or if you move into an executive position, the groundwork you've laid in EQ – like listening before making decisions, or staying calm and constructive during crises – will support you in high-pressure strategic roles. In short, treating emotional intelligence as a lifelong development journey means you'll never stop improving your "people skills," which keeps you effective and relevant at every career stage. It's just like updating your technical toolkit – except the tool you're sharpening is yourself.

EQ Beyond the Workplace

Emotional intelligence isn't something you check at the office door. On the contrary, the benefits of high EQ extend far beyond the workplace – into your personal relationships, your well-being, and your daily happiness. Engineers may be logical thinkers, but we're also friends, partners, parents, and community members. Applying EQ skills in these roles can greatly enrich your life. Think about it: the same active listening you practice with a colleague can make a huge difference when talking with your spouse or a close friend. The empathy that helps you understand a client's perspective can also help you connect with your children. And the self-regulation techniques that reduce your stress during a product launch can equally help you stay calm when life throws you a curveball at home.

Stronger relationships through empathy and listening. One of the most practical ways EQ shows up in personal life is in our conversations with loved ones. How often have we engineers been accused of being "distant" or too solution-focused when a family member just wants us to listen? By consciously using *active listening* – giving someone our full attention, acknowledging their feelings, and resisting the urge to jump straight into problem-solving – we can become better partners and friends. For example, imagine your friend or partner has had a bad day and is venting. Instead of the stereotypical engineer response ("Let's fix this by doing X..."), an emotionally intelligent approach would be to listen patiently, say *"I hear you – that sounds really tough,"* and maybe ask *"How are you feeling about it now?"* This simple act of empathy can dramatically improve your connection. The other person feels understood rather than brushed aside. People high in EQ excel at such interactions – research shows they navigate everyday interpersonal situations more smoothly and maintain healthier relationships as a result. In fact, emotional intelligence has been linked to greater relationship satisfaction overall. By practicing empathy and clear communication at home, you can prevent small misunderstandings from escalating and build a foundation of trust and respect with those you care about.

Reducing stress and enhancing well-being. Work can be stressful, and many engineers know the feeling of bringing that stress home – maybe you've snapped at a family member after a long day, or lost sleep ruminating about a bug in the code. Strengthening your EQ can help break that cycle. Skills like self-awareness and self-regulation are essentially your stress management tools. When you're self-aware, you notice *"I'm feeling frustrated and tense right now"* rather than unknowingly letting that mood dictate your behavior. That awareness gives you a chance to pause and use coping strategies (deep breathing, a walk, listening to music – whatever helps you reset) before you inadvertently take it out on others. Emotional intelligence also encourages a positive outlook and resilience. Instead of catastrophizing a mistake ("I messed up this project; my career is over"), a high-EQ person reframes it ("I made an error, but I can learn from it and recover"). This healthier thinking reduces anxiety and helps you bounce back faster. Over time, these habits profoundly affect your well-being. Studies have found that people with higher emotional intelligence experience lower levels of stress and more frequent positive moods – in other words, they tend to be happier and healthier. They also cope with challenges in more constructive ways (for example, discussing a problem openly instead of withdrawing or resorting to unhealthy habits). By cultivating EQ, you're not just becoming a better coworker – you're also doing your mental health a favor. Many engineers report that learning to manage their emotions (rather than ignore or suppress them) led to reduced burnout and a better work-life balance. You might find you're able to "leave work at work" more effectively, or handle conflicts at home with the same thoughtful approach you apply on the job.

Becoming a better partner, parent, and community member. Emotional intelligence, at its core, is about understanding and managing emotions – both your own and others'. These skills are incredibly valuable in family and community life. For instance, consider parenting: children don't come with an instruction manual, and logical reasoning alone often fails when a toddler is throwing a tantrum or a teenager is feeling misunderstood. Parents with high EQ use empathy to validate their child's feelings ("I see you're upset because your toy broke – I'd be upset too") and self-control to remain patient instead of yelling. This doesn't mean being a pushover; it means responding in a calm, supportive way that teaches the child how to handle emotions constructively. Over time, such emotionally intelligent parenting can foster trust and emotional security in the child. Similarly, as a partner in an adult relationship, EQ helps you navigate the inevitable disagreements or stressful times. Rather than reacting with anger or shutting down, you can communicate openly about how you feel and listen to your partner's perspective. You might say, *"I felt hurt by what happened at dinner. Can we talk about it?"* – a response that invites dialogue instead of argument. It takes courage and practice, but it leads to stronger, more honest relationships. And the benefit isn't just to others; it reflects back on you, creating a supportive environment where *you* feel understood as well. No wonder research shows a clear link between emotional intelligence and relationship quality. High-EQ individuals tend to enjoy more satisfying and harmonious personal relationships, both at home and in friendships.

Beyond family life, emotional intelligence can also make you a better community member and leader outside of work. If you volunteer in your community or participate in any group (from a local coding meetup to a sports team), your EQ skills help you collaborate and earn trust. You might be the one who mediates when there's a disagreement in the homeowners' association, or who shows kindness to a struggling neighbor. These "soft" gestures have a real impact. By practicing empathy, fairness, and good listening in your community, you set a positive example and often inspire others to do the same. In essence, *practicing EQ holistically – not just in the office – turns you into a more well-rounded, compassionate human being.* The reward is a richer life: deeper friendships, a happier home, and the knowledge that you're contributing positively to your community. And here's an encouraging thought: the interpersonal skills you sharpen outside of work can loop back and make you more effective at work too. When you learn patience by helping your kid with homework, you might find that patience comes in handy when mentoring a junior developer. When you learn to read the mood of a room at a community meeting, you might use that insight in the next big team meeting at the office. In this way, growing your EQ beyond the workplace creates a virtuous cycle, improving all realms of life simultaneously.

Balancing IQ and EQ for Success

As an engineer, you pride yourself on your IQ – your analytical intelligence, logical reasoning, and technical expertise. These "hard skills" are undeniably important; they help you solve complex problems and build great products. But technical brilliance alone is not enough for maximum success. Time and again, both research and real-world experience show that combining IQ with EQ – in other words, leveraging both your brainpower *and* your people skills – creates a far more potent and well-rounded skill set. The most effective engineers (and leaders) are those who use their intelligence to come up with solutions, and their emotional intelligence to ensure those solutions are adopted, teams function smoothly, and stakeholders are on board. In a sense, IQ might get an idea on the table, but EQ is what gets that idea across the finish line.

Consider a scenario: Two engineers propose competing designs for a system architecture. Engineer A has a technically superior design on paper, but they dismiss feedback, struggle to explain the benefits in plain language, and get irritated when colleagues ask questions. Engineer B's design is solid though maybe not as fancy; they actively listen to concerns, adapt the design based on team input, and generate enthusiasm by acknowledging everyone's contributions. Which design is more likely to be implemented successfully? In many organizations, Engineer B's blend of technical and emotional intelligence will win the day. That's because the best idea in the world won't make an impact if you can't communicate it, collaborate on it, and get buy-in from others. Technical skills (IQ) and interpersonal skills (EQ) work best in tandem. As one industry expert put it, the top engineers "know how to combine their technical proficiencies with the *soft skills*" that lead to better results – including stronger relationships, effective decision-making, and even higher creativity on teams.

Research supports this balance. Notably, studies have found that cognitive intelligence (IQ) by itself accounts for only about 25% of career success – the rest comes down to other factors, like emotional intelligence. In fact, in workplace settings people with average IQs often outperform those with the highest IQs, which was long a puzzle to scientists until EQ emerged as the "missing link" explanation. High-EQ individuals excel in teamwork, adaptability, and leadership, which boosts their performance and advancement. It's no surprise then that 90% of top performers score high in emotional intelligence. Technical brilliance might land you a job, but EQ is what helps you *excel* in that job and rise to leadership. Many employers have taken note of this: a survey found 71% of employers value EQ more than IQ in employees, because they know how crucial "people skills" are on the job. Think about that – your ability to manage emotions and connect with others could be even more important to your employer than your raw technical acuity. That doesn't diminish the value of your engineering know-how; it simply underscores that without EQ, your IQ can't shine fully.

True success comes from a synergy of hard skills and human skills. This is evident in team dynamics and project outcomes. A highly analytical engineer might devise a brilliant solution to a problem, but if they lack EQ, they may fail to convince decision-makers or alienate their team in the process (limiting the solution's impact). Conversely, an engineer with moderate technical ideas but excellent EQ might rally a team to accomplish something great together, improving upon the idea as they go. Ideally, of course, you want both strong technical chops and strong emotional skills – that's a powerful combination. For example, a software architect with top-notch coding abilities and the empathy to understand user needs is going to design more user-friendly, successful products than one who only focuses on the code in isolation. As another illustration, an engineering department might be filled with PhDs, but if they can't collaborate or if morale is low, their output will suffer. There's a saying in management: "IQ gets you hired, but EQ gets you promoted." It rings true in many cases. The engineer who can troubleshoot a system and mediate a team argument is management material. The developer who can write efficient code and mentor junior colleagues is bound for leadership.

Even companies known for their analytical rigor have discovered the supreme importance of EQ in practice. A famous example comes from Google, a company founded on cutting-edge computer science. Google initiated a data-driven study of what made managers effective (Project Oxygen), expecting technical expertise to be a top factor. The results were eye-opening: technical knowledge ranked dead last among the traits of Google's best managers. Instead, the most valued qualities were things like good communication, listening, empathy, and mentoring – classic EQ skills. In other words, even in one of the world's most elite engineering environments, human-centric skills trumped technical prowess for leadership success. As Google's VP of People Operations noted, they realized a great manager isn't the smartest engineer in the room, but rather someone who *"made that connection and was accessible"* to their team. This doesn't mean technical skills don't matter – Google hires brilliant minds, of course – but it shows that once you meet a baseline of technical ability, it's your emotional intelligence that differentiates you. Blending technical and people skills creates a more productive, innovative workplace, because team members feel heard, supported, and motivated. Projects led with a balance of IQ and EQ tend to hit their targets more reliably, since the team is both competent *and* cohesive.

So, as you progress in your career, remember that you don't have to choose between being a brainy engineer or a people-oriented one. The best engineers leverage both sides: they solve complex problems *and* bring out the best in others. If you're naturally strong in IQ, challenge yourself to grow in EQ areas like listening or collaboration. If you're a people person, make sure you're also continuously learning technically. By balancing and strengthening both, you become the kind of well-rounded professional who can lead projects and inspire teams to success. Technical expertise might get a project started, but emotional intelligence is what keeps it going and ensures it delivers impact.

Future-Proofing Your Career

The engineering landscape is evolving faster than ever. Technologies like artificial intelligence and automation are changing the nature of work, and some tasks that once required lots of brainpower can now be done by machines. In this context, the uniquely human abilities – empathy, creativity, collaboration, leadership – are becoming the true differentiators for future engineers. Emotional intelligence is not just a nice-to-have; it's what will future-proof your career in an increasingly logical, digital world. While coding languages or tools may come and go, human connection and insight will always be in demand. Cultivating your EQ is one of the best long-term investments you can make in your professional resilience.

Think about the rise of AI: algorithms can crunch numbers, diagnose problems, even write code to an extent. But what they *can't* do (at least not yet) is genuinely understand human emotions, build trust, or create innovative ideas out of heartfelt inspiration. As routine technical tasks become automated, engineers will spend more time on roles that require human touch – like guiding strategy, interfacing with clients, brainstorming creative solutions, and leading diverse teams. In these areas, emotional intelligence is the superpower that sets you apart from any robot or algorithm. A recent insight in the industry put it plainly: *emotional intelligence remains a unique human skill, critical for career growth in 2025 and beyond.* Being able to understand context, navigate relationships, and make judgment calls is where humans have the edge. In other words, your EQ is what makes you adaptable and valuable, even as technical skills are augmented by AI. This is why the World Economic Forum listed emotional intelligence among the top skills needed in the workforce of the future and projects that it will remain in high demand for years to come. The logic is simple: the more we incorporate advanced tech into work, the more our human traits stand out as key assets.

Emotional intelligence also makes you adaptable and resilient amid rapid change. In the tech world, the one constant is change – whether it's a new programming paradigm, an unexpected market shift, or a sudden pivot in project goals. Engineers with high EQ are typically better at weathering these changes. Why? For one, self-awareness and self-regulation help you manage the stress and uncertainty that come with change. Rather than panicking or resisting, you're able to stay calm, assess your emotions, and maintain a positive outlook that *"we can figure this out."* This resilience is crucial when learning new skills or switching roles becomes necessary. Additionally, EQ includes the quality of adaptability itself – being flexible and open rather than rigid. People with strong EQ tend to embrace learning opportunities and adjust their approach when circumstances shift, instead of clinging to "the way we've always done it." For example, if a new automation tool comes along that disrupts your usual workflow, an adaptable, emotionally intelligent engineer will be curious to explore it (rather than threatened by it) and empathetic in helping colleagues transition. They might say, *"This is a big change for all of us. Let's see how we can support each other in mastering this new system,"* showing leadership through the transition. In contrast, someone low in EQ might either ignore the change (to their detriment) or handle it insensitively, causing more friction. EQ equips you with the mindset to continuously evolve, which is the essence of staying relevant in any field transformed by technology.

Top industry leaders have been vocal about the importance of retaining our human empathy in the age of AI. One prominent tech CEO gave a commencement speech where he remarked that *he isn't worried about computers learning to think like humans – he's more worried about people thinking like computers, without values or compassion.* This striking comment highlights the concern that as we dive deeper into technology, we must not lose our human touch. We don't want engineers to become unfeeling logic machines; on the contrary, the more tech we have, the more crucial it is that the people behind the tech are grounded in empathy and ethics. When he said this, he was urging the next generation of engineers to carry their humanity with them, to ensure technology serves people's needs and well-being. Similarly, another tech leader has said that "empathy is not a soft skill… it's the hardest skill we learn", underscoring that truly understanding and relating to others is challenging but absolutely essential. These insights from visionaries drive home a clear point: the future will belong to those engineers who pair their technical acumen with emotional wisdom. The algorithms may handle the computations, but humans must guide how technology interfaces with society – and that guidance requires emotional intelligence in spades.

In practical terms, imagine the engineer of the future: Perhaps she's leading a project to integrate AI into healthcare devices. Her IQ helps her understand the complex algorithms, but her EQ is what allows her to empathize with end-users (patients and doctors), to collaborate with a multidisciplinary team, and to anticipate ethical dilemmas. When technical issues arise, she uses her EQ to keep the team motivated and focused rather than in panic. When users give feedback, she listens and adjusts the product to better suit human concerns. This combination makes her incredibly effective. The human-centered skills of creativity, empathy, and teamwork become the X-factor that no machine can replicate. It's telling that even as automation increases, many jobs of the future are expected to emphasize interpersonal interaction, innovation, and leadership – areas where emotional intelligence is key. By investing in your EQ now, you are essentially future-proofing your career: you'll be prepared not only to adapt to new technologies, but to take on the roles that machines can't, such as visionary team leader, empathetic client advisor, or change management champion.

Finally, strong EQ will poise you for leadership opportunities that emerging technologies create. As routine tasks are automated, engineers often have a chance to step up to more complex responsibilities – coordinating projects, interfacing between technical and non-technical groups, or driving strategy. Those with high EQ will find it easier to step into these leadership roles because they can manage relationships and inspire others. Emotional intelligence is frequently what separates a great individual contributor from a great leader. As the logical tasks become easier for AI, the *emotional* tasks – like motivating a team or negotiating stakeholder buy-in – become even more defining of success. The good news is that by honing your EQ, you're not only becoming a better colleague today, you're also ensuring you have the skills needed to thrive tomorrow. One could say that in a world increasingly run by algorithms, emotional intelligence is the human superpower that will keep engineers indispensable and at the forefront of innovation. By mastering EQ in this logical world, you position yourself to not just survive the future, but to lead and flourish in it – all while keeping the very things that make us human at the heart of technology.

In summary, lifelong growth in EQ is your secret weapon as an engineer. Commit to continuous learning in emotional intelligence as fervently as you do in technical domains. Apply your EQ skills broadly – at work, at home, everywhere – to enrich your life and others'. Balance your brilliant mind with a big heart, because true success requires both. And as the future unfolds, rest assured that your capacity for empathy, adaptability, and human connection will set you apart in a logical world. The journey of EQ mastery is ongoing, but with each step, you're becoming not just a better engineer, but a better leader, friend, and human being. That is the ultimate achievement in any career.

Epilogue

The blueprint for emotional intelligence exists within every engineer's toolkit; we simply need to recognize that human connection follows patterns as predictable and learnable as any algorithm. Throughout these pages, we have explored how the same analytical minds that solve complex technical challenges can master the intricacies of emotion, empathy, and interpersonal dynamics.

Engineering has always been about building bridges: spanning rivers, connecting networks, linking ideas to reality. Emotional intelligence represents perhaps the most crucial bridge of all—the one that connects brilliant technical minds to the hearts and motivations of colleagues, clients, and communities. When engineers embrace this connection, extraordinary things happen.

The most innovative teams emerge when logical problem-solving meets emotional awareness. The most successful projects launch when technical excellence partners with human understanding. The most transformative leaders rise when engineering precision combines with genuine empathy. These are not opposing forces competing for space in our professional lives; they are complementary strengths that amplify each other.

Consider the engineer who can decode both system failures and team frustrations with equal skill. Picture the project manager who troubleshoots technical bugs while simultaneously addressing communication breakdowns. Imagine the technical lead who designs elegant solutions while fostering an environment where every team member feels valued and heard.

This dual mastery represents the future of engineering leadership. As our technological landscape grows increasingly complex, the engineers who thrive will be those who can navigate both the logical and emotional dimensions of their work with equal fluency.

The journey toward enhanced emotional intelligence begins with a single step: acknowledging that understanding people requires the same curiosity and systematic approach we apply to understanding systems. From that foundation, every interaction becomes an opportunity to practice, refine, and strengthen these essential skills.

The engineering profession stands poised to lead not just in technical innovation, but in creating more collaborative, empathetic, and ultimately more human workplaces. The tools are in your hands.